D1049722

# MEN, WOMEN, AND THE BIRTHING OF MODERN SCIENCE

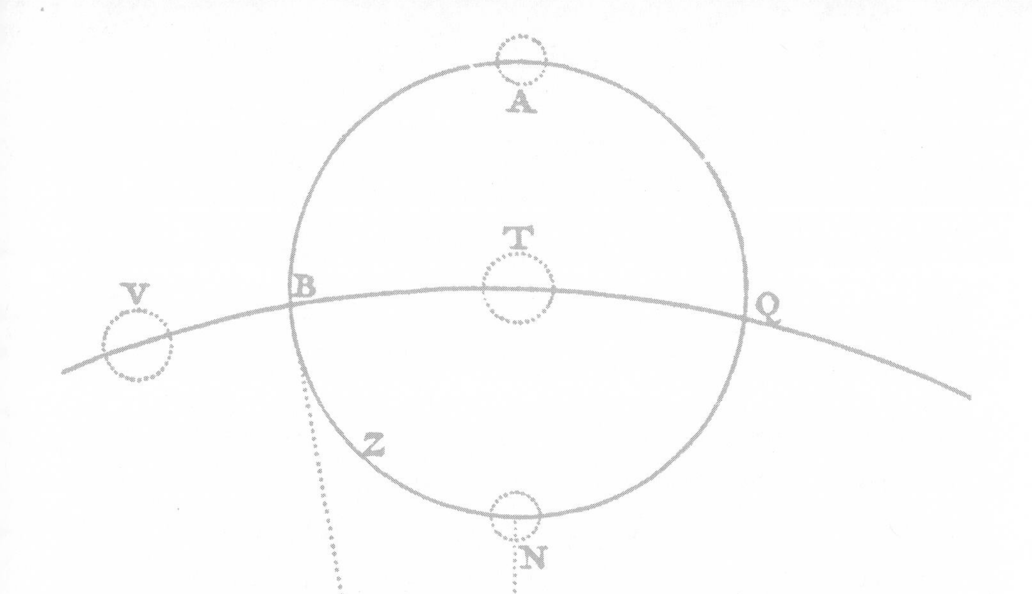


# Men, Women, and the Birthing of
# Modern Science

*Edited by* Judith P. Zinsser

Northern Illinois University Press / *DeKalb*

Published by the Northern Illinois University Press, DeKalb, Illinois 60115

Manufactured in the United States using acid-free paper



Design by Julia Fauci

Library of Congress Cataloging-in-Publication Data

Men, women, and the birthing of modern science / edited by Judith P. Zinsser.

p. cm.

Includes bibliographical references and index.

ISBN-13: 978-0-87580-340-1 (clothbound : alk. paper)

ISBN-10: 0-87580-340-7 (alk. paper)

1. Feminism and science. 2. Sexism in science—Europe—History. 3. Science—

Social aspects—Europe. 4. Science—Europe—History. 5. Science—Philosophy.

I. Zinsser, Judith P.

Q130.M46 2005

500'.82'094—dc22

2004030000

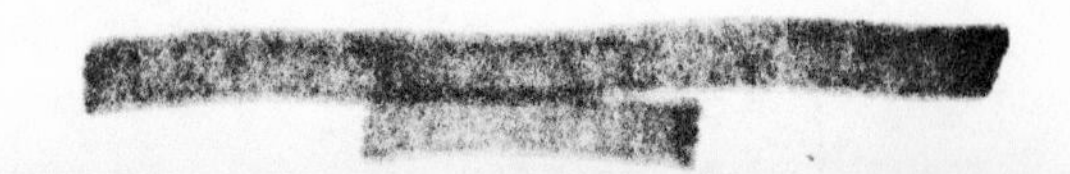

# Contents

# Acknowledgments

This collection of essays, *Men, Women, and the Birthing of Modern Science,* arose from a three-day conference sponsored by Miami University and the University of Cincinnati. As a result, there are many institutional sponsors to thank for bringing these scholars together. I am grateful to the International Visiting Scholar Exchanges Fund; the Havighurst Fund; the Sigma Chi/William P. Huffman Scholars in Residence Fund; the McClellan Symposium Fund; the departments of History, French, and Italian; and the programs in Interdisciplinary Studies and Women's Studies at Miami University; the Charles Phelps Taft Conference Fund at the University of Cincinnati; and the offices of the Dean of Arts and Sciences, the graduate schools, and the provosts of both universities. Special thanks are owed to the chairs of the respective history departments, Charlotte N. Goldy and Barbara N. Ramusack.

At Miami, Marty Miller in Special Collections, Pamela Salela and Susan Hocker at King Library, and Edna C. Southard of the University Art Museum created special exhibitions that complemented the conference sessions.

Other participants in the conference enriched the discussions, challenged traditional assumptions, and went on to further study of the gendered nature of early modern natural philosophy in other gatherings and in their own publications. These include Paula Findlen, Jan Golinski, Sandra Harding, Mary Terrall, Maria-Teresa Medici, Brita Rang, Willard Sunderland, and the evaluators Brian W. Ogilvie and William H. Schneider. The editor owes particular thanks to those who remained part of the collection for their advice, encouragement, and patience with what proved to be a far lengthier process than any of us imagined.

At Northern Illinois University Press, acquisitions editor Melody Herr shepherded the project through the acceptance process with exceptional skill and understanding. Her suggestions were always constructive and thoughtful. Kelly Parker, the editor who worked on the manuscript, intuited what authors meant to say and smoothed over the different publishing and historiographical conventions evident in such an international collection. It is a much better volume because of her queries and changes.

A number of colleagues at Miami University kept the project going and solved last minute problems: Mary Frederickson, Stephen Norris, and Anna Klowoska. Jeri Schaner transformed long e-mail attachments and numerous corrections into a manuscript and thus made the collection a reality. Elizabeth Smith, as always, gave her competent assistance.

The collection would not exist without the initial encouragement and the continuing enthusiasm of Hilda L. Smith at the University of Cincinnati. We planned the original conference together; she raised half the funding; and her considerable experience with this kind of enterprise proved invaluable at every stage of the endeavor.

For any book, there are always those for whom there is no specific attribution; rather they are individuals whose contributions are impossible to define in a few words. Therefore, Murray D. List, Sarah K. Lippmann, and Roger J. Millar also must be mentioned.

# MEN, WOMEN, AND THE BIRTHING OF MODERN SCIENCE

JUDITH P. ZINSSER

# Introduction

The birthing of modern science was a long and circuitous process. Nothing about our modern understanding of the term, its methods, or its practitioners evolved simply or easily. During the three centuries of its evolution, much was discarded, much was lost, and many were excluded from its study. If Galileo, who lived from 1564 to 1642, is claimed by traditional historians of science as one of the first "scientists," his use of the telescope for his observations of sun spots and his experiments with falling bodies represented only the beginning of the process of definition. Over a hundred years later in 1778, the newest edition of the official dictionary of the Académie française, the institutional arbiter of the French language, defined "science" as "knowledge that one has of things." The only aspect of the term that even remotely corresponded to our very specific, modern understanding of the discipline was the assumption that this science was the result of study and would be more certain than other kinds of knowledge. In the seventeenth and eighteenth centuries one could speak of the "science of good and evil," and metaphysics, despite its ties to what we today would call "theology," was considered the science of beings in the abstract.

As the essays in this collection demonstrate, study of the motion of objects led even the most practical experimenter to questions about the nature of God's role in the universe. If there was a law governing the impact between particles, believed to be the basic components of all matter, what did this mean about God's ability to act once they had been set in motion? If human beings were only clusters of colliding particles, what of the soul, free will, and the choice to sin or to be virtuous? Only gradually, over the course of the seventeenth, eighteenth, and nineteenth centuries, did the concept of science lose its broad, general connotation and its associations with philosophy and questions of religion. Not until late in the nineteenth century did it acquire its familiar specific meaning: the study of nature based on mathematics, observation, and controlled experimentation.

In addition, throughout this long transition, the manner of study, the manner of validation, and the qualifications of practitioners remained malleable. The old authorities, such as the faculties of the universities, found their knowledge and their methods of reasoning questioned. Only gradually, however, did new institutions evolve and gain their own authority. In these circumstances women of the privileged classes, some of whom appear in essays in this collection, had an opportunity to engage in the intellectual discourses of their day. Christina, the seventeenth-century queen of Sweden, was only one of a number of women who not only corresponded and debated with Descartes about his separation of the soul from the material body but also conducted their own scientific inquiries. Queen Christina was interested in alchemy, the beginnings of our chemistry. Her contemporary in England, Margaret Cavendish, Duchess of Newcastle, wrote about the microscope, the nature of the universe, and the position of women in the law, to name only a few of the subjects that engaged her considerable energies. In the eighteenth century, women such as the marquise Du Châtelet in France and the women of the Winkelmann-Kirch family in Germany made their contributions to scientific discussions and enterprises.

Not abruptly, but gradually over the last decades of the eighteenth century and the beginning of the nineteenth century, the definition of science narrowed and then closed. "Scientific method" came to mean particular techniques requiring particular training, while mathematical descriptions of the universe came to be acknowledged as more exact models of the observed world. Questions about causes, beyond what man could learn and his methods verify, were laid aside. Over time, for example, Newton's law of attraction was accepted even though there was no verifiable explanation of *why* it functioned as it did; the mathematics and their correlation with observed phenomena merely illustrated *how* it worked. "Certain" knowledge validated by mathematics, by direct observation, and by repeated experimental demonstration emerged as "science," the discipline that we know in the twenty-first century.

During this shift, the metaphorical language of experimental science remained neutral. "Nature," for example, could be both female and male. The same was not true for those who studied and described nature. Popular images of women studying the universe and the new mathematical science shifted and portrayed frivolous, flirtatious amateurs. The definition of "feminine" traits that had existed for centuries prevailed, as did the characterization of the female as physically incapable of rational thought and subject to the unpredictability of her reproductive organs and emotions. Women's activities and their contributions came to be mocked and discredited, ignored and forgotten, or, at best, subsumed by those of the men they worked with. New institutional, intellectual authorities, such as the Royal

Societies and learned journals, had emerged in many European capitals, and women, who at first had actively participated in scientific activities and speculation, found themselves marginalized.

This collection, then, describes this era of transition. Beginning with women "scientists" in early modern Europe, and following the evolving definition and consequent gendering of science and its practitioners, the essays bridge the gap between histories of the scientific discoveries and controversies of the sixteenth and seventeenth centuries and the histories of modern science in the nineteenth and twentieth centuries.

## Natural Philosophy, Gender, and the Institutionalization of Science

When in 1605 Francis Bacon in his great treatise *The Advancement of Learning* envisaged a way to order all knowledge, it seemed a sensible and possible endeavor. Almost 150 years later, the mathematician Jean le Rond d'Alembert would refer to Bacon's "Tree of Knowledge" when he came to construct his own categorization of all that man knew for the *Encyclopédie,* the great project of the French Enlightenment. At the pinnacle of both of these hierarchical representations was the study of "natural philosophy." To write this phrase now, in the twenty-first century, this conjunction of the scientific observation of nature with the speculative reasonings of philosophy invites confusion.

For those of us raised in the era of space exploration and computers, it seems hard to believe that science, that most exact study of the natural world, could once have had anything to do with the abstract realm of philosophy. But in early modern Europe the results of observation, experimental verification, and mathematical models functioned within the same framework as the concepts of rhetoric, epistemology, and metaphysics. As Margaret J. Osler explains in her contribution to this collection, studies of the natural world demonstrated the majesty of God's creativity and power and the ways in which humanity might know and understand these divine phenomena. Equally hard to believe is that the gradual process over the course of the eighteenth and nineteenth centuries by which "natural philosophy" became "science" meant choices that not only separated out the study of God's role in the universe but also determined the basic model for *all* knowledge in our modern secular world.

Much has been written about the end of this story in the nineteenth century, and particularly about Darwin's theory of evolution and its theological repercussions. But Darwin's ideas would never have been accepted had not a series of choices already been made about what constituted valid knowledge, what could be known "scientifically," and who would make these decisions. In the seventeenth and eighteenth centuries, all of these choices were contested: what to call "fact," what constituted a reliable experiment, and who could be considered credible witnesses. Also in the process of formulation were decisions about who might participate in these choices and the institutions that could validate them and thus give authority to the men and women who had made them.

"Natural philosophy" gradually became narrowed and divided until, by the middle of the nineteenth century, the term "science" described a particular kind of inquiry, with its presumption of fact, objectivity, and neutrality based on trained observation, witnessed experiment, and mathematical models. Thus delineated, "science" referred to the world of nature as it can be known by man through the use of reason and the senses: in short, verifiable truth. "Philosophy" in contrast, became that part of knowledge concerned with first causes, with the ways we know, and with the relationship between the knower and the observed, and between the knower and the causes of what is observed. In the course of the nineteenth century, this type of knowledge fell away from the apparent certainties of science. Philosophy, and its constituent domains such as aesthetics and metaphysics, became associated with "speculation"—all that could not be described or verified by mathematical formulas, experiments, and trained observation.

The phrases "Scientific Revolution" and "Enlightenment," which usually designate this era of transition, were coined in the 1930s by intellectual historians to describe the thinking and writing of an informal, elite community of individuals who functioned largely in opposition to or outside of the traditional universities and circles of the learned. Members of this community corresponded with each other, visited back and forth across the Continent from London to St. Petersburg, met in Academies, and gathered in salons and coffeehouses to discuss natural philosophy. Though men made up the vast majority engaged in these discussions and speculations, no aspect of the study inherently excluded women. Indeed, women participated in these studies and discoveries. Despite Western Europe's centuries-old traditional denigration of women's intellectual capacities and women's automatic exclusion from academic institutions, women furthered the evolution of the new knowledge.

Gradually, however, the old attitudes, revitalized by the language of experiment and observation and their unquestioning acceptance by the newly legitimated scientific elite, were once again used to "prove" even the brightest and most privileged females, both by their nature and their physiology, were unsuited to either scientific or philosophic inquiry. Women's access to specialized training diminished. Following the practice of the church-related universities before them, the academies, societies, and other sites of experimentation were reserved for men. By the end of the eighteenth century, women performed roles deemed useful, but secondary, to the work of their male mentors or patrons. Any ideas or discoveries were attributed to their male colleagues, dismissed as the work of amateurs, or viewed as marginal and easily forgotten. This subtle combination of traditional attitudes and changed circumstances gave a gendered dimension to both the new science and the new philosophy.

This collection of essays explores this transitional period, the development of these two concurrent aspects of seventeenth- and eighteenth-century intellectual history: (1) the gradual separation of what the learned of early modern Europe called "natural philosophy" into what we now categorize as "science" and "philosophy"; (2) the participation of women despite the gradual reaffirmation of that knowledge as a masculine domain, its practitioners as male, and access to its new institutions as the prerogative of men. Ten international scholars have written on these themes from the perspectives of England, France, Germany, Italy, Sweden, and Russia. Together, the essays offer the opportunity not only for specific national study but also for a cross-cultural analysis of these two important aspects of the history of early modern Europe. Thus, this collection is unique: first, it offers insights from the history of science *and* the history of women in the seventeenth and eighteenth centuries; second, it surveys these two themes across Europe.

## Description of the Collection

The collection is divided into three sections. Section one, "Women Natural Philosophers," offers descriptions of three elite women who participated in this eclectic study in the seventeenth and eighteenth centuries. Susanna Åkerman in "Queen Christina's Metamorphosis" explains that the monarch presented herself as an equal to the learned men of her day in both metaphysics and natural philosophy. She argued theology with Protestant and Catholic churchmen and established her own Academy in Rome after her abdication from the Swedish throne. She explored the transformative powers of alchemy with the idea of changing her body from female to male. Margaret Cavendish, Duchess of Newcastle, as described in Hilda L. Smith's "Margaret Cavendish and the Microscope as Play," insisted on her intellectual superiority, presenting her ideas on natural philosophy in many genres, from plays and novels to poems and treatises. She often took an independent position, in this case rejecting the use of artificial instruments like the microscope to study the universe. Cavendish, like Emilie de Breteuil, the subject of Judith P. Zinsser's "The Many Representations of the Marquise Du Châtelet," used her writings to gain acceptance from male contemporaries. Du Châtelet also took unorthodox positions. Her major work, the *Institutions de physique* (Institutions of Physics), published in 1740, was an original synthesis of Descartes' method, Leibnizian metaphysics, and Newtonian mechanics. The marquise's concern for validation led her to represent herself equally adept as philosopher, mathematician, and physicist.

Section two of the collection, "Shifting Language, Shifting Roles," presents analyses of writings by seventeenth- and eighteenth-century male natural philosophers in England, France, and Italy and illustrates the ways in which Robert Boyle's gender-neutral language of the seventeenth century was superceded by the gender-rich images of eighteenth-century authors. Margaret J. Osler's essay "The Gender of Nature and the Nature of Gender" demonstrates the continuing significance of theology and the relative insignificance of gender in Robert Boyle's explorations of the natural world. In contrast, J. B. Shank's "Neither Natural Philosophy, Nor Science, Nor Literature" and Franco Arato's "Minerva and Venus" describe how concepts of masculine and feminine and appropriate male and female roles were central to Bernard le Bovier de Fontenelle's description of Descartes' cosmology and Francesco Algarotti's description of Newton's natural philosophy. Fontenelle and Algarotti wrote their scientific dialogues in a style suitable for the salon. Shank argues that Fontenelle chose this form and style to highlight the importance of imagination, as well as reason, a quality more often associated with literature than natural philosophy. Algarotti expected his work to teach and to entertain. Both of their portrayals of a learned mentor and his eager young woman pupil validated the possibility of female participation, but, at the same time, their references to the rituals and pleasures of flirtation and seduction could be used by others to suggest more traditional images of women's frivolity and sexuality.

The essays in the last section, "Women, Men, and the New Scientific Establishment," offer examples of the rhetorical and practical ways in which women's contributions were marginalized, denigrated, or subsumed by learned men. Yet they also demonstrate women's strategies for continuing to participate in scientific activities despite narrowed definitions, professionalization, and the increasing primacy of exclusively male institutions for its study. Many of the seventeenth-century Englishwomen of Lynette Hunter's "Women and Science in the Sixteenth and Seventeenth Centuries" were related to the founders of England's Royal Society. They experimented with and developed substances for cooking and medicinal purposes in ways later enshrined as "scientific" when performed by men in spaces men had identified as appropriate for such work. However, these women's experiments fell away, denigrated as "kitchin-physick." Similarly, as described in Stephen Clucas's "Joanna Stephens's Medicine and the Experimental Philosophy," a woman's formulation for dissolving kidney stones was first praised, then appropriated, and finally dismissed by the male professionals of the British Royal College of Physicians and the learned men of the Royal Society. Monika Mommertz in "The Invisible Economy of Science" demonstrates how the women of the Winkelmann-Kirch family refused to accede to the changed circumstances of this institutionalization and masculinization of the new science. Though officially barred from the Berlin Academy and its observatory and never credited for their work, they continued to collect the astronomical and astrological information that provided much of the Academy's income and its European reputation. They supported themselves by this "household production" of knowledge and thus remained part of a scientific enterprise now in theory reserved to men.

The eighteenth-century Russian princess E. R. Dashkova was a visible symbol of women's competence in all kinds of learning. She used her position as supervisor of both St. Petersburg Academy and the Russian Academy of Sciences to support translations of Western writings, including *On the Nobility and Advantage of the Female Sex,* a humanist tract favoring a positive view of women's nature and capabilities. Yet, as Grigory A. Tishkin explains in his essay, Dashkova's appointment was by imperial fiat and thus dependent on the continued favor of Catherine the Great. Dashkova's role in the publication of an antimisogynist tract remained anonymous, and its message proved ephemeral amidst the more traditional images presented in even the most daring of the contemporary eighteenth- and nineteenth-century periodicals for women.

Whereas Europe's women continued to play significant roles in the late eighteenth-century world of the new science, the strategic choices available to them often obscured their activities. In addition, their contributions were often subject to the needs of those in authority, the new professionals of the reconfigured learned establishment. Neither science nor philosophy, as these fields became demarcated and their study and practice formalized, would be considered appropriate endeavors for women. The regendering of learning had been achieved.

Join us then in reading about these fascinating times that shaped our views of knowledge, our belief in science and its methods, and the challenges faced by women with a love of learning.

# SECTION I:

# WOMEN NATURAL PHILOSOPHERS

SUSANNA ÅKERMAN

# Queen Christina's Metamorphosis

## *Her Alchemical World Soul and Fictional Gender Transformation*

The reason for Queen Christina of Sweden's abdication in 1654 was generally held to be her refusal to marry. Also her choice to convert to Catholicism in 1656 was seen as a result of her early attraction to the Catholic ideal of virginity.[1] At the same time underground pamphlets told of a scandalous romance with her Spanish advisor, Antonio Pimentel, and more pointedly with an unnamed Jewish woman in Hamburg.[2] Her letters in these years to "la belle," the courtier Ebba Sparre, and to the wife of the French Ambassador, Charlotte de Brégy, suggest physical attraction, and it was reported to Rome that Christina had "none of the woman except the sex."[3] Controversy has continued over the sexual persona of the Queen, but it is clear that she saw herself as having a male character and intellectual interests. She wrote that she had "a hot and dry temperament," thus following the trend of her astrologers that used this description for her constitution, actually the male condition in Galen's humoral pathology.[4]

Christina styled herself as the Convert of the Age, and she set up court in Rome, where she held a series of scientific and cultural academies in her palace. Her Accademia Reale was staged briefly in the Palazzo Farnese in her first year in Rome, 1656, but was revived in 1674 and was held for a number of years in her own Palazzo Riario. In addition, Giovanni Ciampini's Accademia dell'Esperienze, also called Accademia fisico-mathematico, met in her palace for the first time in 1677. She was protectress of the Accademia degli Stravaganti in Collegio Clementina from 1678 and in Orvieto of the Accademia dei Misti.[5] Her inspiring presence and resources were valued by many literary men. After her death in 1689, she was chosen symbolical figurehead, Basilissa, by the poets that formed the Accademia d'Arcadia.[6]

Although Christina has a place in modern philosophy books because of her invitation to Descartes in 1646, it is in fact very difficult to discern her philosophical opinions.[7] She did not write any sustained philosophical document through which we can learn of her ideas. She wrote maxims and a few letters with questions to Descartes, but neither of these are very revealing sources for her philosophical ideas.[8] On the other hand, there are many reports of her conversations that include references to her having intellectual opinions in a philosophical context. There is ample evidence that in her searching period, before her conversion to Catholicism, she read intensely in philosophical and religious material. It is symptomatic that often she was reputed at court to take an absurd or extreme stand in debates in order to test the opponent.[9] Testifying to her wide reading are the Jesuits' statements based on their secret meetings with her in 1651, reporting that Christina had searched among the views of the ancients, of the Jews, and of the heretics before making her decision to convert to Catholicism.[10]

Although Christina started out her meetings with Descartes during his visit to Stockholm in 1650 by reading his newly finished exposition *The Principles of Philosophy* (1644), Descartes in fact had to admit that Christina's interest in reading Greek manuscripts, through the help of scholars such as Isaac Vossius and Claude Saumaise, left her little time to read and discuss his own philosophy. Christina did not like his advice that she should abandon these Greek readings and responded that she had already read his opinions in the Greek skeptic Sextus Empiricus and in Plato.[11] By contrast, Christina is described as "trembling with joy" when she received a copy of Iamblichus' *De Mysteriis aegyptiorum, chaldeorum et assyriorum,* describing methods of theurgy and divination, the ascent of the soul, and how to come in contact with gods and daemons.[12] Although Christina later said that Descartes gave her "the first lights" on Catholicism, his philosophy appears really not to have influenced her.[13] There is only one echo of Descartes in her maxims, the statement that "Before one can come to believe, one must doubt."[14] From the correspondence of Isaac Vossius we learn instead that Christina read the Neoplatonists: Olympiodorus, Proclus, and Hermias and their commentaries on dialogues of Plato, such as Olympiodorus on the *Phaedo* and the *Philebus*. From Paris she wanted to obtain Proclus' commentaries on the *Alcibiades* and the *Parmenides*. Vossius says that Christina developed this interest from an early reading of the Florentine Platonist and Christian kabbalist Pico della Mirandola.[15] Vossius does not indicate which work of Pico she read, perhaps the *Dignity of Man* or the *Heptaplus,* but it is clear that she was impressed by his humanist view of the ancient philosophers, that they argued for a concordance of principles, and that a single universal doctrine lay behind appearances, the Platonic vision.[16]

It is significant that before the Jesuits' arrival, whom she specifically asked to be skilled in mathematics, Christina was in contact with the Roman Jesuit and polymath Athanasius Kircher. In 1649 Kircher had presented her with a copy of his *Musurgia Universalis* (1650) on the harmony of music and the soul of the world.[17] Kircher also spoke briefly of his work on the *Oedipus Aegyptiacus* (1652–54), which told of an ancient revelation of Egyptian-Hermetic mysticism, Kabbalah, and oriental magic. Kircher in his Hermetic fervour even addressed her in November 1651 as "Regina serenissima, potentissima, sapientissima, vere trismegisii," thus suggesting to her that she contemplate further the three spiritual crowns of Hermes Trismegistus.[18] Significantly, Christina mentioned in her response to Kircher that her letter was carried by Macedo,[19] the Jesuit who first encountered Christina's interest in Catholicism. Macedo arrived in disguise in Stockholm in 1650 in the suite of the Portuguese Ambassador. In September 1651 he reported to the Jesuit General Goswin Nickel that Christina wanted to have secret talks on Catholicism.[20] Paolo Casati, Kircher's colleague at the Collegio Romano, and Francesco Malines were chosen for the task. The connection to Kircher, and his first letter to Christina in June 1649, point to the philosophical background of a perennial philosophy showing the ascent of the soul to the One that the Jesuits probably addressed theologically with the young Queen. They may have emphasized Dionysius the Areopagite and his Christian and Neoplatonic vision of a celestial hierarchy of angelic beings, but perhaps not, for it is only later that Christina is mentioned reading his texts.[21] To complicate the narrative of her conversion, Christina also wanted to obtain Porphyry's tract *Against the Christians* that Vossius claimed was available in the Bibliotheca Laurenziana in Florence.[22] In 1652 Christina also suggested to the Greek scholar Johannes Schefferus that he write a history of the Pythagoreans. Schefferus completed this task a decade later when he published the first Pythagorean history in early modern Europe, his *De natura et constitutione scholae Italicae sive pythagoricae* (Uppsala, 1664).

Even if little is known of Christina's early interest in these ancient cults and philosophies, we have a witness to her special opinion on immortality during her later life in Rome. This witness is the natural philosopher Gottfried Wilhelm Leibniz, who wrote an essay in which he discussed Christina's philosophy. His report shows that Christina's Platonic philosophy developed into a very radical point of view. Leibniz was very interested in meeting Christina personally, and during his Italian journey in 1689 he went to Rome to see her and the Vatican library.[23] At first, he was angered to hear that Christina had fallen ill, the illness from which she would die in the same year. Later, however, he wrote: "I thought the Queen was dead . . . but she is out of danger and I could see her."[24] In 1702 Leibniz wrote an essay for the use of Queen Sophie Charlotte of Brandenburg and chose the theme of the doctrine of a single universal spirit.[25] In this essay, he mentioned what he had heard about Queen Christina's philosophical opinion on immortality. The essay began:

> Some discerning people have believed and still believe today, that there is only one single spirit, which is universal and animates the whole universe in all its parts, each according to its structure and the organs which it finds there, just as the same wind current causes different organ pipes to give off different sounds. Thus they also hold that, when an animal has sound organs, this spirit produces the effect of a particular soul in it but that, when the organs are corrupted, this particular soul reduces to nothing or returns so to speak, to the ocean of the universal spirit.[26]

Leibniz provided the medieval source for this early modern belief:

> Aristotle has seemed to some to have had an opinion approaching this, which was later revived by Averroes, a celebrated Arabian philosopher. He believed that there is an *intellectus agens,* or an active understanding, in us and also an *intellectus patiens,* or a passive understanding, and that the former . . . is eternal and universal for all, while the passive understanding, being particular for each, disappears at man's death. This was the opinion of certain peripatetics two or three centuries ago, such as Pomponatius, Contarini and others, and one recognizes traces of it in the late Mr. Naudé, as his letters and his recently printed *Naudeana* show.[27]

Leibniz pointed out that these Averroistic debates led to the assertion that the doctrine was true according to philosophy, but that it was false according to faith. This finally resulted in the fierce disputes concerning the twofold truth, a doctrine condemned by the Lateran Council.

Next Leibniz made his startling claim: "I have been told that Queen Christina held a strong inclination towards this opinion, and since Mr. Naudé, her librarian, was saturated with it, it would seem that he gave her information about these secret opinions of famous philosophers, with whom he had discoursed in Italy."[28] Thus, Leibniz portrayed Christina as having a radical, and ultimately heretical, Averroistic opinion—the view that there is no personal immortality and that the world is eternal. This view contradicted traditional Christian doctrine and did not agree with the usual picture that the Church promoted of Queen Christina as a Catholic champion. However, the belief in a World Soul as a solution to the question of life after death can be seen as an intellectual viewpoint that Christina adopted alongside her religious beliefs. She may have been convinced that the possibility of the abstract joining of the active intellect with the World Soul was a more reasonable view of the afterlife than the concrete images of bliss and damnation, purgatory and judgment, taught and defended by the Church.

Leibniz continued by making comparisons:

> Spinoza, who recognizes only one single universal substance, is not far from the doctrine of a single universal spirit and even the neo-Cartesians, who hold that only God acts, affirm it, seemingly unawares. It would seem that Molinos

> and certain other modern quietists, among them an author who calls himself John Angelus Silesius, who wrote before Molinos . . . and even before Weigel, shared this opinion of a Sabbath or repose of souls in God. It is for this reason that they believed that the cessation of particular activities is the highest state of perfection.[29]

Leibniz thus posited a connection both to Christina's early life in Stockholm, when Gabriel Naudé visited the court and took care of Christina's library, and to the later period in Rome in the 1680s when Christina was interested in Miguel Molinos' mysticism.[30] The Quietists held that by an inward turning towards the Divine in which all active thoughts were silenced one would reach a perfect stillness and union with the Divine. Concrete images of devotion were abandoned in favour of an intellectualized purification of the conception of the Divine. Leibniz's claim appears to have been that the Quietism of Molinos could be grounded in an Aristotelian-Averroist view of the Soul through which union with the universal spirit is made possible. To set all these World Soul believers on a par with Averroes, however, seems to tie too many aspects in one bundle. But Leibniz may have been correct in seeing metaphysical and practical similarities in such a wide range of ideas. In 1681 the Quietists were condemned by the Church and Molinos was imprisoned, accused of sexual abuse and heresy. It is said that Molinos' letters to Christina, over two hundred of them, were removed from the investigation out of respect for her. Others say she herself had them burnt.[31]

Leibniz traces the Averroist view to Gabriel Naudé's influence over Christina. But perhaps we should be skeptical here and take account of the fact that Naudé was in Stockholm for only two years, 1652–53, and that he left Stockholm in anger. He claimed that Christina had lost all interest in books and that she was not forthcoming with his salary.[32] However, it is interesting to note that Christina did own several manuscripts written by the Averroist Renaissance Aristotelians mentioned by Naudé: *De incantationibus* (Ms. Reg. Lat. 1280) and *De substantia orbis* (Ms. Reg. Lat. 1279) by Pomponatius, or Pietro Pomponazzi, and commentaries on Aristotle's Physics (Mss. Reg. Lat. 1154–55) and *De Anima* (Ms. Reg. Lat. 1280) by Cesare Cremonini.[33] Her library, now deposited in the Vatican, the Codices Reginensis Latini et Graeci, is very rich, numbering approximately 2,200 Latin, mostly medieval, texts and 150 Greek manuscript items.

In addition, there are other documents that attribute a World Soul philosophy to Christina that are of an early date and that must be discussed in this context. In particular, there is the controversial document of 1655, the anonymous tract entitled *Le Genie de la reine Christine de Suède,* that straightforwardly claimed that Christina believed that there was no immortality except through a union with Plato's World Soul.[34] This tract was probably designed to destroy Christina's Catholic reputation at the time of her official conversion at Innsbruck and her travel to Rome. The tract is very interesting, however, as it outlines the essential points of a convinced freethinker critical of the objective forms of religion. The tract claims that Christina held five philosophical principles, opinions that, in fact, have some relation to Christina's known interests at the time of her conversion and that are characteristic of her rather elitist view of religious affairs.

The principles were said to be: (1) One should love God, not fear him. We should instead fear the vile men and real devils in whom common people have such faith. (2) One must follow one's principles and seek no repentance or pardon. (3) Religion at present is but a poor illusion. Certain particular sentiments are good and sound, while the common ones are full of error. (4) There are no good arguments to refute the doctrine of the universal soul, and there is no immortality except that after death the soul returns to its principle, Plato's universal soul of the world. (5) Moses was an impostor as well as a bon ésprit—a good wit. Christina is said to have openly doubted the biblical story of the miraculous passage of Moses through the Red Sea.[35]

It is important to note that views such as these may have been known to Christina. She was, for instance, in 1652 prepared to read a copy of the infamous tract *De tribus impostoribus (On the Three Impostors),* a tract that argued that Jesus, Moses, and Mohammed had designed religion in order to gain political power. Christina's advisor Johan Adler Salvius was said to have a copy, but he threw it on the fire before Pierre Bourdelot (later one of her mentors) could grab it.[36] The tract *De tribus impostoribus* had been discussed in letters of scholars ever since the mid-sixteenth century. Yet it seems that the copies in Latin with this title that are now found in various libraries throughout Europe were only written in the 1680s, probably by someone who wanted to capitalize on the interest in clandestine books and especially those which were critical of religion.[37]

Christina's belief in a World Soul may be compared with the report of what Christina said later that year, in 1656 in Dijon on her way to Paris from Rome and thus after her official reception by the Vatican. She answered a question put by three scholars and recorded by J. B. Morisot, author of an alchemical text of praise for the solar-King set in the land of the golden Inkas, *Peruviana* (1644). She was asked what her true religion was. Characteristically, Christina answered that her religion was that of the philosophers. "This philosophy is indeterminate, and its limits are uncertain, but it is best represented in Lucretius *De rerum natura.*"[38] Lucretius described an atomistic world governed by huge masses of matter in motion, atoms that in a continuing flow compose and decompose in infinite, but ultimately random, sequences. In this world there is neither fate nor Divine order. Yet we may infer that Christina was probably also influenced by the contemporary spread of Pierre Gassendi's atomism, for she had invited Gassendi to Stockholm after the death of Descartes in 1650.[39] Gassendi attempted to Christianize the atomistic world system in order to make the essential composition and decomposition of atoms less a matter of mere randomness.

Through Pierre Bourdelot, Christina was able to gain a copy of Gassendi's *Animadversiones in decimum librum Diogenes Laertii,* a lengthy treatise in which Gassendi argued that if the number of atoms in the world were not infinite, but limited, God could project his providence onto the world and thus control the ordering of atoms.[40] But as Gassendi never came to Stockholm, it appears that her view of this atomism was influenced by her physician, Bourdelot, an avant-garde intellectual whose arguments often tended towards atheism. Bourdelot was a broad-minded and amusing person and was appointed to Christina's court in 1652–53. He is known to have advised Christina to stop her obsessive reading in order to cure her melancholy. But one should note that he also gave Christina a number of manuscripts on the universal monarchy written by the condemned heretic Tomaso Campanella, whom he had personally known in Paris where Campanella had come after he was released from prison in Italy.[41] Bourdelot's uncle, Jean Bourdelot, had published an extensive bibliography of Pomponazzi's work in 1633, so it might well have been with Pierre that Christina could first discuss Averroism.[42] After Christina had settled in Rome, Bourdelot continued to inform her of events in the intellectual world of France and in 1664 sent her a copy of an atomistic philosophy written by Gilles de Launay. She did show some interest in mechanistic philosophy when Alfonso Borelli's *De motu animalium* (Rome, 1681) was dedicated to her, but her personal preferences lay in Neoplatonism.[43]

It must be remembered that when Christina was planning her secret conversion ceremony, which took place at Antwerp in the Spanish Netherlands in 1655, she was at the same time actively projecting an image of being an atheist. She may have done this because it was better for her to be regarded as an atheist than as a Catholic convert, probably in order not to anger the Swedes who were then debating her revenues.[44] Thus, one solution to the question of her anomalous beliefs is to say that Christina was in fact arguing for principles similar to those presented in the tract *Le Genie de la reine Christine de Suède,* if only to deceive her visitors. On the other hand, Leibniz spoke of her belief in a World Soul. Christina may have held such opinions even in her later period in Rome and especially in her Quietist period that began in the mid-1670s.

Leibniz described the belief as contemporary with other early modern views. He said:

> In itself the doctrine is good, for all who teach it recognize in fact the existence of divinity, whether they believe that this universal spirit is supreme—in which case they hold that it itself is God—or whether they believe, like the Cabalists, that God created it. The latter is also the opinion of the Englishman Henry More, the Cambridge Platonist, and other newer philosophers, particularly of certain chemists who believe that there is a universal Archeus or world-soul; some of them have maintained that this is the spirit of the Lord moving over the waters, of which the beginning of Genesis speaks.[45]

Here one can see a link between Christina's interest in alchemy and the alchemist's belief in an active operative force within matter, the Archeus, which can be assimilated to the World Soul, the location for eternal ideas and for forms of perfection.

However, Leibniz also criticized the World Soul as being based on a false, imaginary notion. It is not enough merely to have an imaginary notion of it based only upon a very lame comparison with wind-animating musical organs. This analogy even might encourage the idea that the wind changed the pipes and thus support ideas of the preexistence of the soul or transmigration of souls from one body to another.[46] Leibniz set this image of musical organs and pipes beside another image, one of active intellect merging with the World Soul in the same way a drop merges in the great ocean of water. He wrote:

> . . . if one imagines the universal spirit to be like an ocean composed of an infinity of drops which become detached when they animate some particular organic body but are reunited to the ocean after the organs are destroyed this leads to a false view of the Divine. Since the ocean is an aggregate of drops, God would be an assemblage of all souls, almost in the same way that a swarm of bees is an assemblage of small animals.[47]

This would lead to a Divinity without an essential unity.

Thus, Leibniz dismissed the World Soul, but it is clear that he had laid out its structure as belonging to different schools of natural philosophers, among them Averroists, Platonists, kabbalists, and alchemists. That Christina could have held such a belief is perhaps strengthened by the broad-based content of the notion of the World Soul and its convincing basic parallel with how a drop of water joins the great ocean of water. There may have been great comfort for her in this view of life after death. Christina probably saw the World Soul as part of an elitist, esoteric, and secret belief fitting her role as an intellectual outsider at the same time as she sought protection from the Catholic Church.

The alchemical side of the World Soul was part of her inspiration. There is no evidence to determine exactly when Christina started with alchemy, but her involvement tended to increase toward the end of her life. Her first acquaintance with alchemy may have been when in 1651 in Stockholm she was approached by the alchemist Johannes Franck. He described her future reign as the fulfillment of Paracelsus' prophecy of a return of Elias Artista and of Sendivogius' vision of the rise of a metallic monarchy of the North.[48] With these visions Franck urged the Queen to start searching for the ruby red powder of the philosophers. He expressed these hopes in the tract that he offered her: *Colloquium philosophicum cum diis montanis* (Uppsala, 1651).[49] A year later, there were reports that an Italian alchemist by the name of Bandini had arrived at her court.[50] Clear evidence of the Queen's own practise of alchemy appears first when the Danish alchemist

Olaus Borrichius met her in Rome in 1665. He reported that he had often talked with her about the study of chemical arcana and experiments and that she, as a "Palladio virago," dedicated herself entirely to the sacred art.[51] In the summer of 1667 in Hamburg, Christina experimented with the messianic prophet and alchemist Giuseppe Francesco Borri, but Cardinal Azzolino wrote her that she had to distance herself from Borri because he was being sought by the Inquisition.[52] Christina at this time also corresponded with Johan Rudolf Glauber, the discoverer of Sal mirabile, or sodium sulphate (Na2SO4), who claimed that this discovery was in fulfillment of Paracelsus' prophecy of Elias Artista, the accomplished alchemist who would appear at the end of time but who now appeared in the form of a salt, Et Artis salia. Christina posed seventeen apparently simple questions to him on the nature of the various alchemical stages, such as: what colour the material is when it is reduced to its ultimate perfection; whether it is subject to change; whether it is affected by liquors; what doses to take of various ingredients and their colours in the process, etc.[53] She also took an interest in the phosphorus discovered by Hennig Brandt.[54] In her collection of spiritual medieval manuscripts there are some forty alchemical manuscripts by the foremost medieval authors, as well as practical handbooks. They included works by Geber, Johan Scotus, Arnold de Villa Nova, Raimund Lull, Albertus Magnus, Thomas Aquinas, Bernard Trevisano, George Ripley, George Anrach d'Argentine, Johan Grasshof, and the intriguingly illustrated *Rosarium Philosophorum*—with its alchemical image of the merging of the solar-King and the lunar-Queen into a hermaphroditic union.[55]

Significantly, Christina owned a version of Massimiliano Palombara's alchemical manuscript *La Bugia* (Ms. Reg. Lat 1521) and was implicated in his raising of the notorious Porta Magica in his gardens on the Esquilline hill in 1680. Palombara briefly mentions an Order of the Golden and Rosy Cross in his own version of *La Bugia* (The Little Candlestick) from 1656.[56] The signs on the Porta commemorate the alchemical process and are topped by an emblem taken from Henricus Madathanus' allegory *Aureum Seculum Redivivum* (1622) allegedly written by a golden cross brother, an *"aureae crucis frater."*[57] There is also a contemporary French manuscript in Christina's collection called *Veritas Hermetica veritati quearenti seu de differentia inter Chymicam nostri temporis et antiquam* (Ms. Reg. Lat. 1218) that sheds new light on the alchemy of the Rosicrucians. This text expounded on the gathering of dew and its processing in the preparation of the stone and referred to some Fratres Roris Cocti—brothers of boiled dew—an alchemical way of referring to the Rosicrucian fraternity.[58] The text explained how a crystalline fluid that could be gained from supercelestial water, called *"maim"* by the Hebrews (as in *"aesch majim,"* the fiery water or watery fire seen in the turquoise blue colours of the sky). The anonymous author referred to a speech (in Theophraste Renaudot's *Bureau d'adresse* in Paris) on 16 May 1639 "before an assembly of *'la Rose-Croix'*" explaining the importance of dew: "The true menstruum of the Red Dragon [the true matter of the philosophers] knowledge of which this society has wanted to pass on to posterity through a name whose marks cannot be erased by time, keeping the name Brothers of Boiled Dew *[Frères de la Rosée Cuitte]* . . . the blessing of Isaac and Jacob did contain but two matters, the dew of heaven and the fatness of the earth *[de Rore caeli et pinguedine terrae]*." The last phrase translates as "God give thee of the dew of heaven and the fatness of the earth" after Isaac's blessing of Jacob in Genesis 27:28. The phrase can also be found on the title page of John Dee's *Monas Hieroglyphica* (Antwerp, 1564) with the exhortation "Let the water above the heavens fall and the earth will yield its fruit."[59]

There has been much speculation as to why Dee's astro-alchemical Monas sign, Mercury with Aries (the Ram) below, is added by Johann Valentin Andreae to his *Chymische Hochzeit des Christiani Rosenkreutz, Anno 1459* (Strassburg, 1614).[60] Christina could confer with the source directly as she owned a handwritten German copy of Dee's book (Ms. Reg. Lat. 1266). It is bound in red moroccan leather with her Vasa family crest printed in gold on its covers. Her manuscript *Veritas Hermetica* essentially explained the inner alchemical belief of the Rosicrucians and their dependence on Dee's philosophy, that coction of dew and its empowering by light is the crucial step in the alchemical opus, i.e., *ros crux*. The alchemical process is further brought to light in a tract dedicated to Christina by Giovanni Batista Comastri, the *Specchio della Verità, concordanza sopra la filosofia hermetica* (Venice, 1683). Comastri wrote that the philosophical solvent *(solve et coagula)* is found in the air, hinting at the extraction of *Sal nitre,* i.e., saltpetre or potassium nitrate (KNO3), from dew in the tradition of the Cosmopolite, i.e., from the pathbreaking *Sal nitre* theory in *Novum lumen chemicum* (1608) published by the Polish adept Michael Sendivogius. Comastri borrowed his phrase: "There is in air an occult bread of life, the congealed spirit of which is better than the whole world *[universa terra]*."[61] This chemical vapour or soul pervades all of matter, and Sendivogius' extraction of it has been seen to prefigure the discovery of oxygen.[62] The dew that contains this salt, however, is for Dee, since he knew Postel, more akin to the life-giving substance described by the Jewish Kabbalah, as in the *Zohar*'s commentary on the above phrase in Genesis 27:28, where dew flows "from the brain of the Ancient of days" and is understood as an erotic psychosexual presence, as the supercelestial watery fire flows in.[63]

Christina owned several manuscripts that undergird this worldview, such as a copy of the Hermetic *Asclepius* ending with a prophecy that after dark ages of ignorance the Egyptian religion will be restored. She also possessed the Florentine Hermeticist Marsilio Ficino's commentaries on Plato's *Parmenides* and on Iamblichus' theurgy—calling down the angels. Her collection included Trithemius' angelic cryptography *Steganographia* (Ms. Reg. Lat. 1344) from which Dee sought inspiration for his Monas design. This volume also contains Al-Magritti's *Picatrix,* describing the preparation of magical talismans and the utopian city Adocentyn. Founded by Hermes

Trismegistus, the city is ruled by a priest-king in perfect harmony with the cosmos. Other unusual items in her collection were parts of a *Libro de las formas en de las ymagenas que son en los cielos* (Ms. Reg. Lat. 1283) and a Latin version of the *Sefer-ha-Raziel* (Ms. Reg. Lat. 1300), books of angelic magic from the thirteenth-century court of Alfonso X (the Wise) of Castile.[64] Adding to these were texts on natural magic by the Renaissance Aristotelians Lefevre d'Etaples and Pietro di Abano (Ms. Reg. Lat. 1115), as well as numerous copies of the original works of Aristotle and Plato. A copy of the clandestine dialogue among representatives of seven religions, *Colloquium Heptaplomeres,* on the secrets of the sublime, which has been described as a vision of the resolution of religions in a superior harmony, is classified as Ms. Reg. Lat. 1313.[65]

Christina's collection in Rome of printed books numbered over four thousand items and included, apart from a large number of books on pure medicine, the collected works of Paracelsus; alchemical tracts by Joseph Quercetanus, Bernard Trevisano, Martin Ruland, Isaac Hollandus, Oswald Croll, and Andreas Libavius; and volume 3 of the collection *Theatrum Chemicum* published by Zetzner in 1602. The Paracelsian Gerard Dorn's *Clavis philosophia chymisticae* (Herborn, 1599) and Heinrich Noll's *Naturae sanctuarium quod est Physica hermeticae* (Francfort, 1619) contribute to the mystical philosophy of nature also represented in her library by the English esotericist Robert Fludd's *Microcosmi historia* (Part II, Oppenheim, 1617). She also owned the Hermetic *Pimander* in Annibale Roselli's multivolume edition from Cracow (1585). In her section of Hebraica there are more than a hundred titles, including the kabbalist source *Sefer Jezirah* and parts of the *Zohar,* Isaac Abarbanel's messianic prophecies, and kabbalist-messianic commentaries by Moses Almonsini, Menahem Recanati, Rabbi Aquiba, and Salomon Molkho. Also included are texts by kabbalists such as Bahya ben Asher, Shem Tob Ibn Gaon, Abraham Gallico, Salomon Alkahetz, and Menachem Azariah. In addition, Christina had a copy of Menasseh ben Israel's *Nishmat Hayim* (Amsterdam, 1652), defending the transmigration of souls and relating stories of spiritual possession, the dibbukim.[66] In 1655 she gave a large collection of alchemical manuscripts from Prague to her librarian Isaac Vossius. Once owned by Rudolph II and written in the German, Czech, and Latin languages, this collection now resides as the Codices Vossiani Chymici at the University of Leiden.[67]

Christina desired to know more of alchemy and brought a younger woman called Sibylla into the experiments.[68] In 1670 she employed a working alchemist, Pietro Antonio Bandiera, to run her laboratory, and she ultimately willed the equipment to him.[69] Her own practise in alchemy has left only a few traces on paper. There exist a drawing of alchemical equipment with short comments written by Christina on calcination and a query on how many hours the fire should burn.[70] There is a document in her own hand, entitled "Il laboratorio filosofico—paradossi chimici," which is either her own framing of a presentation of alchemy or the notes from a text with the same title (see appendix).[71] She was preoccupied with alchemy to the very end of her life; found by her deathbed in 1689 was a letter by Samuel Forberger on the universal medicine, Glauber's alkahest, and secret fire.[72]

Was Christina perhaps an adept involved in the Roman alchemists' closest circles, or was she merely a patron prepared to do alchemy by proxy through her servants? She clearly was a very forceful woman. She claimed that her mind was entirely masculine and that she lacked what she saw as the faults of womanhood, such as a will to subordination. This belief was to materialize in her ardent hope for a real transmutation. As Kjell Lekeby has discovered and describes in his controversial study *Kung Kristina: Drottningen som ville byta kön (King Christina: The Queen Who Wanted to Change Sex),* there is in her collection of papers now in Riksarkivet, Stockholm, which she willed to Cardinal Azzolino, an Italian text on which Christina had written that it was given to her on 21 April 1682.[73] In it, Christina's abdication and travel to the great metropolis of the world where she will devote herself to a cult that only she will know was described. It then depicted her life in 1683. Suddenly, one sentence reads, "la natura perfectionerra l'opera," and a strong, brave youth by the name of Alexander appears. The text went on to tell of Alexander's future travel to Constantinople to convert the Turks by appearing in their mosques. He then collects an army and fights a battle at the Nile after which he collects his trophies and returns in triumph to Rome in 1694, where he is crowned "Imperatore dell'Universo" by the Pope, Leone il Massimo. Alexander and Leone are to rule in peace over the whole world in a golden age. Since Christina, as ex-Queen, took the name Christina Alexandra in Rome, it appears that the prophecy with its wonderful metamorphosis spoke to her inner dreams of perfecting herself. In this the Aristotelian view of women as undeveloped men had a role to play, as did the alchemical vision of polarities and ultimate perfection. The prophecy is inspired by the underground Joachimite tradition that flourished in the sixteenth century with visions of a future Universal Monarch and Angelic Pope who would reform the Church from within.[74]

Lekeby builds on research by historian of sexuality Thomas Laqueur showing that in the early modern world the Aristotelian conception of women had led to the widespread belief that they were biologically incomplete versions of men.[75] In principle, however, given the alchemical theory of transmutation and nature's metamorphoses, it was possible for Christina to believe that she could perfect herself and ultimately reach the higher state of complete malehood. Other evidence brought forth by Lekeby supports this idea of sexual transformation and appears to express the social expectations about a ruling female monarch. In 1680 Abbé Serviens wrote to the French minister Colbert that four years before, in 1676, Christina had what we now would call a "prolapsed uterus" and started to believe that she was transforming into a man. A chambermaid had made an inspection and had fallen to her knees at the sight of the outgrowth and exclaimed, "Salve rex suecorum [hail king of the Swedes]."[76] Christina is said to have been disappointed when the mistake was made evident by a doctor three months later. Serviens wrote: "She communicated her doubts to her medical doctor, to her chamber maid Ottavia . . . to her surgeon, to Marquise Pignatelli, to the Jesuit father Pallavicino, who gave her his compliments, and to Cardinal Azzolino. . . . Her aspirations were such that she let herself be painted in armour, with a helmet on her head, the vizier elevated, and with the inscription of only one of her names: Alexander rex Suecorum. . . . I can assure you the truth of these circumstances without any exaggeration, but that I have come to know them exactly as I have been told by the people that I have named."[77] Thus, Serviens assured Colbert that his story was not a disrespectful joke or a rumour based on hearsay. The statement does show how Christina's masculinity appeared to the public, and how it was accepted as the true condition of a transgendered ex-Queen.

Another determining factor in forming that image of her is the theologically androcentric expectations about a Christian heroine, that not until she leaves her role in reproduction and materiality and assumes male reason will she fully participate in Christ.[78]

An illustration of these assumptions can be found on the title page of a long solemn poem in twelve songs from the turn of the seventeenth century, Michael Capellari's *Christinas sive Christina lustrata* (Venice, 1700), in which the Swedish Queen's identification with the male is further glamourized: Christina, dressed as Minerva, ascends to the heavens, leaving debris from war and a sculpture of a naked woman's body strewn on the ground. Christina is praised because she has transcended the female condition. Christina's androcentric thinking is also expressed by her own self-image in her autobiography, where she thanks God that her soul has been made totally masculine ("toute virile, aussi bien que le reste de mon corps"), but also thanks him for being born a woman, for even with her masculine character it was better not to be repeatedly tempted to debauchery in the way men are.[79]

Lekeby compares Christina's desire for a transformation with those of the eighteenth-century Madame d'Urfé de la Rouchefoucauld in Paris, who was visited by the Swedish alchemist Nicholas Oelreich, to whom d'Urfé explained that Casanova was in the process of regenerating her into a young man by giving her a tincture made from the blood of a three-year-old boy to drink.[80] This literal interpretation by women of the possibility of fusing male and female aspects into a new regenerated and perfected union has been little considered by authors who discuss the appeal of alchemy to women. The presence of the feminine in alchemy has been much valued by those historians who criticize the cold male rationalism of modern science. The anima of alchemy seems, however, in Christina's case to conceal a strongly androcentric bias, added to by the male dream of control over life-giving processes through the laboratory. Seldom have the implications of the male alchemical aspects for practising women, of which we know more than a handful, been discussed even if Flamel and his wife Perenelle have recently been written out of history by Claude Gagnon as a seventeenth-century crafted fiction.[81] Lekeby's documentation of Christina's case thus has significance for our evaluation of the uses of the alchemical worldview.

Christina has in the alchemical tradition been seen as a male-female phenomenon, perhaps therefore an adept. Yet there may be reasons to doubt her alchemical expertise. In a letter to Azzolino in Hamburg in March 1667 she wrote of the report of a successful transmutation performed by a Dutch burgher.[82] The learned doctor Helvétius, who formerly had been skeptical towards alchemy, was present and now guaranteed its success. Christina added that with one grain of the projection powder one is able to convert "500 livres" of lead, that is 250 kg, into 24 karats of gold. This was far out of proportion as the tradition taught that the real weight is perhaps one grain to 15 g of gold.[83] The idea that the projection powder could transmute a thousandfold was widespread, but to attain this potency there had to be a multiplication process, which she did not mention. Maybe she came to learn more, especially after meeting Borri and after setting up her own laboratory in Rome. Christina knew some parts of alchemy, we may infer, and we may take evidence from one of her maxims: "Chemistry is a beautiful science. It is the anatomy of nature and the true key to open all treasures. It gives fortune, health, glory and true wisdom to its possessor." She added that while alchemy is "a royal, or furthermore, a divine science," it had fallen into utter disrepute among those who think they know everything but know nothing.[84] True to her World Soul she had a medal struck that she showed to her visitors. It carried a shining sun on one side and this text on the other: "Nec falso, nec alieno [with neither false nor borrowed (light)]." This was how she liked to present herself: as a philosopher-Queen, but represented as a solar-King.

The philosophy involved was not the modern rationalism of Descartes but the age-old *philosophia perennis* and the theory of alchemical transmutation. In 1682 the Quietist François Malval claimed that Christina had abandoned Plato and Trismegistos, and he wanted to point to her new mysticism. Leibniz's account, however, suggests that Christina did not abandon the World Soul, but rather saw it in the light of Pomponazzi and Contarini, as an ocean of souls purged from internal hierarchy.

## Appendix

The following short document in Queen Christina's hand,[85] written during her residence in Rome, is intriguing because some of the crossed-out text seems to indicate that she was composing a treatise or perhaps a speech for her academy (although it seems improbable that one would speak of alchemy in a semipublic meeting of this sort). The chapter headings may, on the other hand, be notes from a work that she was reading. This issue was discussed by the late Arne Wettermark many years ago in meetings and correspondence with Eugène Canseliet, who had also commented on the Porta Magica and Christina's alchemical interest in his *Deux logis alchimiques* (Paris, 1944). Although, or perhaps because, Wettermark was deeply involved in research concerning Christina and alchemy, he never published his final thoughts on the document on the metamorphosis that Lekeby later wrote about.[86] If there is a source for her notes, it is probably by a minor contemporary of Christina and therefore very little known.

***Il laboratorio filosofico***
*Paradossi chimici*

**proemio**

1. della antichita delle arte alchimica
2. della sua simplicita
3. della sua brevita
4. della sua nobilita
5. delli errori [communi che si commenciatore in questi arte—crossed out] che si commutione in essa
6. della poder sulfa di [chimici?]
7. delli fornelli
8. dell fuoco [delli vasi—crossed out]
9. della materia uni ogetto del alchimia
10. delli vasi
11. del studio del Autori
12. tempo
13. predicamenti varii che si lavoranno dalli tre regni Animal, vegetabile e minerali, modi di farsi
14. della medicina universale
15. del l'opera [filosofico—crossed out] grande
16. delle virtue luso e vertu di si ha[?]
17. del Magnete

If this outline is Christina's own it would show that she was knowledgeable in the art of alchemy and prepared to lay out its techniques for an audience, probably a limited one. Wettermark was, after his conversations with Canseliet, prepared to regard her as an advanced practitioner. The notion of the accomplished hermaphrodite played an important role in the thinking of Canseliet, who claimed he had last seen his alchemical master Fulcanelli on a stroll in a park in Spain with a woman dressed in Renaissance clothing.[87]

## Notes

1. In her autobiography she wrote lines that she later crossed out, noting that as a nine-year-old child she came to like Catholicism because of "her love for celibacy." See Sven Stolpe, *Drottning Kristina* (Stockholm: Bonniers, 1960–61), 1:4 (1982 repr. ed., 169, 168–71). Autobiography in Christina's papers in Kungliga Biblioteket, Stockholm, and at Bibliothèque de l'école de medicine de Montpellier, Ms. H 258 (15 vols.) and Ms. H 258 bis. (2 vols.).

2. Kjell Lekeby, *Kung Kristina: Drottningen som ville byta kön* (Stockholm: Vertigo, 2000), 50n105 quoting [Jean Carpentier de Marigny] *Histoire de la vie de reine de Suède: Copie d'un lettre écrite de Bruxelles à la Haye touchant la reine de Suède* (Fribourg, 1667). Lekeby reports many comments by French and other observers on her lesbianism, her activity as a *"tribaille."* Lekeby's view is, however, contradicted by Marie-Louise Rodén, who has found letters from Cardinal Decio Azzolino, Christina's official heir, which she claims show that Christina had a love affair with Azzolino. They first met in 1656 on her arrival in Rome. They wrote daily, and in one letter he addressed her as "Cara." See Marie-Louise Rodén, "Drottning Christina and Kardinal Decio Azzolino: Kärleksbrev från det sista decenniet," *Personhistorisk Tidskrift* (1986), 68–74. See also Rodén's augmented dissertation at Princeton, *Church Politics in Seventeenth-Century Rome: Cardinal Azzolino, Queen Christina of Sweden and the Squadrone Volante* (Stockholm: Almqvist and Wiksell, 2000). This activity may, of course, be seen as a lesbian's conversion to heterosexuality, or to heteromorphic chastity.

3. "Muliebris nihil habet preter sexum," Alexander Manderscheidt to the Curia, 3 January 1655, Ms. Barb. Lat. 6487 f. 125–26, Bibliotheca Apostolica Vaticana. This was only to describe her masculine appearance. See Susanna Åkerman, *Queen Christina and Her Circle: The Transformation of a Philosophical Libertine* (Leiden: Brill, 1991), 104n.

4. Stolpe (1960–61), 1:5 (1982 ed., 56–59). Kjell Lekeby, "Till frågan om Drottnig Kristinas konstitution och äktenskapsvägran:ett astrologiskt-medicinskt manuskript från 1650-talet," *Lychnos* (1987), 27–35.

5. *Christina: Drottning av Sverige,* Nationalmusei utställningskatalog 105 (hereafter cited as NMU) (Stockholm: Nationalmuseum, 1966), 377. Apparently there were eighteen volumes of proceedings placed in the Biblioteca Albani. Only one volume has been identified now Cod. Ottoboni 1744, Biblioteca Apostolica Vaticana. NMU, 378.

6. Cesare D'Onofrio, *Rom val bene un'abiura: Storia romana tra Cristina di Svezia, Piazza del popolo e l'Accademia d'Arcadia* (Rome: Fratello Palombi, 1976).

7. Åkerman (1991), 44–69. It was the French ambassador Pierre Chanut who transmitted Christina's questions and who persuaded Descartes to come to Sweden, evidently to extend French political influence over her.

8. Jean François de Reymond, *La reine et le philosophe: Descartes et Christine de Suède* (Paris: Bibliotheque Nordique 5, Lettres Modernes, 1993).

9. Åkerman (1991), 27.

10. Ibid., 30–31. Casatis report reprinted in Leopold von Ranke, *The History of the Popes: Their Church and State, and Especially of Their Conflicts with Protestantism in*

*the Sixteenth and Seventeenth Century* (London, 1906), 3:351–71; Appendix no. 131.

11. F. F. Blok, *Isaac Vossius and His Circle: His Life until His Farewell to Queen Christina of Sweden 1618–1655* (Groningen: Egbert Forsten, MM, 2000), 316.

12. Åkerman (1991), 96–97; Blok.

13. Åkerman (1991), 55–56; Blok, 301–24. For a close view of the Jesuits' activity and Descartes' role, see also Oskar Garstein, *Rome and the Counter-Reformation in Scandinavia,* vol. 4, *The Age of Gustavus Adolphus and Queen Christina of Sweden 1622–1656* (Leiden: Brill, 1992).

14. Ernst Cassirer, *Descartes: Lehre, Persönlichkeit, Wirkung* (Stockholm: Behman-Fischer Verlag, 1939), 185.

15. Blok, 314, 365n., 315n.

16. Ibid., 365n. Christina also owned the Michael Psellus Byzantine commentary on Plato's doctrine of "psychogonia," the generation of the souls, in Ms. Reg. Gr. 131; note also Psellus on the operations of daemons in Ms. Reg. Gr. 136.

17. Pontifical Gregorian University, Rome. Kircher wrote her for the first time in June 1649, APUG, f. 54r-54v. See also Kircher to Christina, no date, f. 50r-50v, available at the Kircher correspondence project on the web (http://www.Pinakes.org). For Kircher and Christina in Rome, see Paula Findlen, *Possessing Nature: Museums, Collecting and Scientific Culture in Early Modern Italy* (Berkeley: University of California Press, 1994); see also her more recent essay in Lo Sardo's *Il Museo del mondo.*

18. Kircher to Christina, 11 November 1651, APUG 561, f. 52r-52v; cf. Athanasius Kircher's Ms. Vat. Arab. 8, Bibliotheca Apostolica Vaticana, "The Psalms of David, De Templo Hierosolimitano a Salomone constructio."

19. Christina to Kircher, no date, APUG 555, 172r-173v, transcription and translation at 276r-276v and APUG 556, 174r-174v. Kircher replies that Macedo has handed him the letter in APUG f. 50r-50v.

20. Garstein, 543–45; cf. 664.

21. Åkerman (1991), 99n. Alexander Manderscheidt wrote a report on 10 October 1653 listing Christina's readings in the Church Fathers (Arckenholtz [1751], vol. 1, 430n222). Dionysius Areopagita's work on the celestial hierarchy and mystical names are in Christina's Vatican collection as Ms. Reg. Lat. 67 and Ms. Reg. Gr. 30.

22. Åkerman (1991), 97; Blok, 118, 124.

23. André Robinet, ed., *Leibniz Iter Italicum: (mars 1689–mars 1690): la dynamique de la République des Lettres : nombreux textes inédits* (Firenze: Olschki, 1988), 176–77.

24. Ibid. "Je croyait la reine morte, mais elle est hors de danger, ces sont mes voeux sant doute qui est contribuées, et je la pourrait voir."

25. See Leroy E. Loemker, ed., *Philosophical Papers and Letters,* by Gottfried Wilhelm Leibniz (Chicago: University of Chicago Press, 1956), 899–910; cf. Åkerman (1991), 79–84. See also Åkerman, "Queen Christina's Philosophy," *Nouvelles de la Republique des lettres* (1991).

26. Loemker, 899.

27. Ibid., 899–900.

28. Ibid., 900.

29. Ibid.

30. Sven Stolpe, *Från Stoicism till Mystik—Studier i Drottning Kristinas maximer* (Stockholm: Bonniers, 1959).

31. Stolpe (1960–61), vol. 2, chap. 9, 9 (1982 ed., 461). See Stolpe, (1959), 258.

32. Åkerman (1991), 82.

33. See *Les Manuscrits de la Reine de Suède au Vatican. Réedition du catalogue de Montfaucon et cotes actuelles,* Studi et testi 238 (Citta del vaticano: Bibliotheca Apostolica Vaticana, 1964).

34. Åkerman (1991), 311–12.

35. Ibid.

36. Ibid., 32. See also Susanna Åkerman, "Christina Alexandra's Search for Clandestine Manuscripts" in *Jean Bodins "Colloquium Heptaplomeres,"* ed. Günter Gawlick and

Friedrich Niewöhner, Wolfenbüttler Forschungen, Bd. 67 (Wiesbaden: Harrassowitz, 1996), 153–64. See also Åkerman, "Johan Adler Salvius' Questions to Baruch de Castro concerning De tribus impostoribus," in *Heterodoxy, Spinozism and Free Thought in Early Eighteenth-Century Europe,* ed. Silvia Berti et al. (Dordrecht: Kluwer, 1996), 397–423.

37. For the most comprehensive research on the *Tribus,* see Winfried Schröder, *Ursprünge des Atheismus: Untersuchungen zur Metaphysik—und Religionskritik des 17. und 18. Jahrhunderts* (Stuttgart-Bad Cannstatt: Frommann-Holzboog, 1998).

38. Åkerman (1991), 73, 77.

39. Ibid., 73–74; Blok, 320.

40. Åkerman (1991), 73.

41. Ibid., 82. Ms. Reg. Lat 1447, 1145, and 443, the latter being Campanella's *Monarchia Messiae.*

42. Jean Bourdelot, *Pietro Pomponazzi . . . operum mss. et vulgatorum nomenclatur* (Paris, 1633); cf. C. Oliva "Note sull'insegnamento di Pietro Pomponazzi," *Giornale critico della filosofia italiana* 7 (1926): 83ff.

43. Åkerman (1991), 255, 259.

44. Garstein.

45. Loemker, 901.

46. Ibid., 906.

47. Ibid.

48. Carl-Michael Edenborg, ed., *Johannes Franck: alkemiska skrifter* (Stockholm: Philosophiska förlaget, 1992). Franck's dissertation from 1645, *De principiis constitutivis lapidis philosophici, theses hermeticae,* 37–39; cf. Sten Lindroth, *Paracelsismen i Sverige till 1600–talets mitt* (Uppsala: Almqvist and Wiksell, 1943), 307. Other expectations were that the transformation would occur around 1658, because of Paracelsus' prophecy of "58."

49. Edenborg, 43ff.

50. Åkerman (1991), 86.

51. Olaus Borrichius, *Conspectus Scriptorum Chemicorum illustrorum: cui praefixa Historia Vitae ipsius ipso conscripta* (Havniae, 1697). ". . . saepe ad disserandum cum regina Christinâ, de arcanioris Chemiae studio, veritate, experimentis, quibus tum sacris se Palladio virago devorerat." I thank Dr. Michael Srigley, Uppsala, for this information.

52. Carl Bildt, *La reine Christine et le cardinal Azzolino. Lettres inédits* (Paris, 1899), 386–89. Borri focused on St. Anne in an attempt to give divine status to her daughter, the mother of Jesus, and claimed that he, as a "pro-Christ" assisted by the angel St. Michael, would lead the Pope to form the Universal Monarchy after defeating the Turks. He had followers in Milan and Rome. See Salvatore Rotta's article on Borri in *Dizionario bibliografico degli Italiano* 13 (1971): 4–12.

53. Azzolinosamlingen 36, Riksarkivet, Stockholm. NMU, no. 900, 370. Glauber died in 1668.

54. Private communication from Anna Maria Partini, Rome.

55. See Christina's manuscript and book catalogue, Ms. Vat. Lat. 8171 in the Vatican library; see also the register of books made after Christina's death in 1689, Ms. Ottob. Lat. 2543, no. 1320. See Anna Maria Partini, "I codici alchemici del fondo Reginense latino alla Bibliotheca Apostolica Vaticana," *Atti dell' Accademia Tiberiana. Anno accademico 1993–1994. Estratti dei corsi e delle conferenze* (Rome, 1994), 158–64.

56. "un compagnia intitolata della rosea croce o come altro dicono dell' aurea croce," in Mino Gabriele, *Il giardino di Hermes: Massimiliano Palombara alchimista e rosacroce nella Roma del Seicento. (Con la prima edizione del codice autografo della Bugia—1656)* (Rome: Editrice Ianua, 1986), 90. For Christina's version, see Marchese Massimiliano Palombara, *La Bugia: Rime ermetiche e altri scritti. Da un Codice Reginense del sec. XVII,* ed. Anna Maria Partini (Rome: Edizione Mediterranee, 1983).

57. Illustrations in Susanna Åkerman, "Queen Christina of Sweden, the Porta Magica and the Italian Poets of the Golden and Rosy Cross" in Adam Mclean's Alchemical Virtual Library (http://www.levity.com/alchemy/queen christina.html).

58. H. D. Schepelern, ed., *Olai Borrichii itinerarium 1660–1665: The Journal of the Danish Polyhistor Ole Borch* (Leiden: Brill, 1983). Borrichius notes after conversations with the Oxford chemist Peter Stahl 23 April 1664: "F.R.C. non vocandis fratres Roseae Crucis, sed Fratres roris cocti, item illum explicuisse illorum insigne (quod est F.R.+) quod + significet lux per anagramma certum, ac si singulari luce essent illuminati, aut quod luce vel aere ad opus suum uterentur," 365. How LVX is derived from the cross X is explained in the sixteenth theorem of Dee's *Monas*. Borrichius also spoke of his visits in 1664 and the many alchemical experiments staged by Pierre Bourdelot in his Academy in France. Bourdelot was sponsored by the Prince de Condé.

59. *Veritas Hermetica,* f. 37. "la vray menstruée de la Dragon Rouge (la veritable matière de philosophe) duquel cette societé ayant voulu laissé à la posterité dans son noms des marques qui ne peuvent etre efacé par le temps, a retenu celuy de frère de la Rosée cuitte . . . la benediction Isaac et Jacob ne contenu que deux matière de Rore caeli et pinguedine terrae." The author said: "Mayer l'a aussi interpreté de même" (probably the alchemist Michael Maier). Borrichius notes in 1664 the same idea: "Ex ore cujusdam Fr:R:C: Ros est in rebus natur: potentissimum Solis dissolvens non corrosivum sed lux ejus inspissanda est et reddenda corporalis, quae tum cocta artificiose in proprio vase tempore convenienti, verum est menstruum rubri draconis, i.e., Solis i. e. verae materiae philosophicae, unde F.R.C. intelligendum Fratres roris cocti. Hinc in gen: benedictio Jacobi haec fuit: De rore caeli et pinguendine terrae det tibi Deus." See Schepelern, 336.

60. See Susanna Åkerman, "The Use of Kabbalah and Dee's Monas in Johannes Bureus' Rosicrucian Papers," paper presented at the "Jewish Mysticism and Western Esotericism" section of the IAHR congress in Durban in August 2000. See also Åkerman, "New Rosicrucian Light on Dee's Monas," paper presented at the Third International John Dee Colloquium, Aarhus, Denmark, December 2001.

61. Giovanni Batista Comastri, *Specchio della Verita—dedicata alla Regina Cristina di Svezia, Venezia 1683,* ed. Anna Maria Partini (Rome: Edizione Mediterranee, 1989). Comastri mentions Cosmopolita and "che è nell'aria un occulto cibo del vita, lo spirito congelato del quale è migliore che l'universa terra," 63. Partini notes (30n) that a description of Christina's laboratory, with all its various equipment, is listed in a manuscript in the library of Fermo, Ms. 4 D.D.2, Cartella XL, 927n.

62. Zbigniew Szydlo, *Water Which Does Not Wet Hands: The Alchemy of Michael Sendivogius* (London: Polish Academy of Sciences, 1994). See 222–36 on how Sendivogius' mention of Neptune is referred to in a text published in 1691 presenting a sigil shaped as a trident: "la Hieroglyphe de la Societé des Philosophes Inconnus." See also Szydlo's "Michael Sendivogius and the Statuts des Philosophes Inconnus" *Hermetic Journal* (1992), 79–91. This is commented upon by Rafal Prinke, "The Twelfth Adept: Michael Sendivogius in Rudolfine Prague," in *The Rosicrucian Enlightenment Revisited,* ed. John Mathews et al. (Hudson: Lindisfarne Books, 1999), 143–92.

63. On Guillaume Postel, Dee, and the Rosicrucians, see Susanna Åkerman, *Rose Cross over the Baltic: The Spread of Rosicrucianism in Northern Europe* (Leiden: Brill, 1998), 173–95.

64. Cf. Susanna Åkerman, "Queen Christina's Latin Sefer-ha-Raziel Manuscript," in *Judeo-Christian Intellectual Culture in the Seventeenth Century,* ed. A. P. Coudert, S. Hutton, R. H. Popkin, and G. M. Weiner (Dordrecht: Kluwer, 1999).

65. Marion Leathers Kunz, "Structure, Form and Meaning in the Colloquium Heptaplomeres of Jean Bodin," in Gawlick and Niewöhner, *Jean Bodin,* 99–120.

66. BAV, Ms. Vat. Lat. 8171, ff. 74–81; F. 57v, 64, 58, 62, 63 Hebraica at ff. 124–30, kindly identified for me by Professor Elliot Wolfson, New York. Menasseh's text is partly in Latin. For Christina's meeting with him in Antwerp, see David S. Katz, "Menasseh ben Israel's Mission to Queen Christina of Sweden 1651–1655," *Jewish Social Studies* (Winter 1983): xlv, 57–72. In 1653 Father Alexander Manderscheidt says of

Christina: "Pour *l'Hebreu* et *Arabe* elle le sait lire & et du moins elle entend un peu" (Arckenholtz, vol. 2, 430).

67. Petrus Cornelis Boeren, *Codices Vossiani Chymici* (Leiden, 1975). Blok shows that Vossius was not interested in alchemy and wanted to sell the collection but found no buyer. The texts were mostly unillustrated, working copies in Czech and German.

68. Anonymous, *Histoire des intrigues galantes de la reine Christine de Suède et de sa cour, pendant son sejour à Rome* (Amsterdam, 1697), 279. Modern edition of the Italian original, Jeanne Bignami Odier and Giorgio Morelli, eds., *Anonymi dell'600 Istoria degli intrigi galanti della Regina Cristina di Svezia e della sua corte durante il di lei soggiorno a Roma* (Rome: Palombi, 1981).

69. Carl Bildt, *Svenska minnen och märken i Rom* (Stockholm: Norstedt, 1900), 174ff. NMU, no. 901, 370.

70. Azzolinosamlingen 36, Riksarkivet, Stockholm. NMU, no. 897, 369.

71. Ibid., no. 904, 371.

72. NMU, no. 305, 364; NMU, no. 902, 370.

73. The document is signed, but the signature is crossed over, so what remains are the letters "Q...ni." Azzolinosamlingen 36, K 429, mapp astrologica, fol. 1, Riksarkivet, Stockholm, cited by Kjell Lekeby, "Drottning Kristinas gudomliga metamorfos: Från intersexualitet till mansblivande," *Fenix: Tidskrift för humanism* 2 (1997): 80–110, 98n84. See also Lekeby's study of her interest in astrology, *I Lejonets hjärta: Drottning Kristina och stjärntydarna* (Stockholm: Pleiaderna, 2001).

74. The Alexander prophecy resembles the Joachimite visions of the Angelic Pope transmitted by pseudo-Methodius and Beatus Amadeus. Christina owned two versions of the *Vaticinia de summis pontificibus* attributed to Joachim di Fiore (Reg. Lat 576 and 1570). She also had a manuscript copy of Guillaume Postel's *Thrésor des propheties de l'univers* (1547), which relates the prophecy of Beatus Amadeus on the Angelic Pope, Ms. Vat. Lat 8171, f. 344; cf. Marjorie Reeves, ed., *Prophetic Rome in the High Renaissance Period: Essays* (Oxford: Clarendon Press, 1992).

75. Thomas Laqueur, *Making Sex: Body and Gender from the Greeks to Freud* (Cambridge, MA: Harvard University Press, 1990).

76. Lekeby (1997), 96. Letter printed in E. Michaud, *Louis XIV et Innocent XI* (Paris, 1882), 1:571ff. Michaud quotes *Correspondence de Rome* t. CCLXIX f. 218–50 in the Archive of the French Foreign Office.

77. Lekeby (1997), 96.

78. On how this is expressed in hagiographical literature that may have influenced Christina, see Eva Haettner-Aurelius, "The Great Performance: Roles in Queen Christina's Autobiography," 55–66. Christina's autobiography is studied also by Kari Elisabeth Börresen, "Christina's Discourse on God and Humanity," in *Politics and Culture in the Age of Christina*, ed. Marie-Louise Rodén, Suecoromana 4 (Stockholm: Svenska institutet i Rom, 1997), 43–53. This attitude is made clear by Norwegian theologian Kari Elisabeth Börresen's studies in early exegesis and the Church Fathers, for whom only men are created in God's image, "whereas women are theomorphic by becoming male in Christ, thus attaining sonship in salvation," which led to praise of "women who achieved perfect maleness by manly virtue, virtus/andreia" (Börresen in Rodén, 45). See also Börresen's, *Subordination and Equivalence: The Nature and Role of Woman in Augustine and Thomas Aquinas* (Washington, DC: University Press of America, 1991), and K. E. Börresen, ed., *Image of God and Gender Models in Judaeo-Christian Tradition* (Minneapolis, MN: Fortress Press, 1995).

79. Arckenholtz, vol. 3, 23. See also Lekeby (2000), 42; Börresen in Rodén, 45–46.

80. Lekeby (2000), 62, builds on evidence from Casanova's Memoirs reported in Carl-Michael Edenborg, *Gull och mull: den monstruöse Gustav Bonde* (Lund: Ellerströms, 1997), 176–78.

81. Claude Gagnon, *Nicolas Flamel sous investigation* (Quebec: Editions le Loup de Gouttière, 1994).

82. Bildt (1900), 174ff.; NMU, no. 898, 369.

83. Lekeby (2001), 71n.

84. Arne Wettermark, "Christine de Suède et la science des rois: Quelques maximes à la lumière de la tradition hermétique," *Nouvelles de la Republique des lettres* 2 (1990): 61–82, 65ff.

85. Azzolinosamlingen 36, Riksarkivet, Stockholm. NMU, no. 904, 371.

86. According to Lekeby, Wettermark had thirteen volumes of material concerning the Queen. See also Arne Wettermark, "Christine de Suède, Roi par la grâce de Dieu," *La Tourbe des Philosophes: Revue d'etudes alchimiques,* 27.

87. Eugène Canseliet, *Deux logis alchimique: En marge de la science et de l'histoire* (1946; repr., Paris: Pauvert, 1979). Fulcanelli (pseudonym), *Le mystère des cathédrales et l'interprétation esotérique des symboles hermétiques du grand ouvres* (Paris, 1974); see also his *Les demeures philosophales* (Paris, 1979).

HILDA L. SMITH

# Margaret Cavendish and the Microscope as Play

Margaret Cavendish, Duchess of Newcastle, was born in 1623 and died in 1673. She published in a wider range of genre than did other seventeenth-century women, including natural philosophy, drama, poetry, orations, biography, autobiography, and letters. Her earliest works were published in 1653 and the last in 1668. While she included scientific topics in most of what she wrote, her more specifically scientific works include *Philosophical Fancies* (1653), *Philosophical and Physical Opinions* (1655), *Nature's Pictures Drawn by Fancies Pencil to the Life* (1656), *Philosophical Letters, or, Modest Reflections upon some Opinions in Natural Philosophy* (1664), *Observations upon Experimental Philosophy* (1666), and *Grounds of Natural Philosophy* (1668). All except the last were reprinted, and a number were corrections of earlier works or contained materials that had appeared before. One could say that as an intellectual, Cavendish was a work in progress, always redoing her writing to overcome earlier criticism.[1]

Cavendish was a royalist and a feminist whose husband, William, was a general, a patron of writers and scholars, and a poet and playwright himself. As an intellectual and social eccentric, she received much attention during her lifetime; gained admiration from the essayist Charles Lamb during the nineteenth century; and had early twentieth-century historians reprint and praise the biography she wrote of her husband. She has been seen, and continues to be seen, in many ways, as an enigma—both as a person and concerning the nature of her intellectual works. Those works were seldom polished, but for many this is their charm. Yet their lack of systematic direction and their variety make it difficult to develop a single assessment of Cavendish as an author. And we have not a great deal more to go on; only a handful of letters remain, and archival sources reflect the perspectives of others. It remains to her works to reveal Margaret Cavendish, and they are often uncooperative in offering up a clear, integrated whole. It often feels as if we have a better sense of her as a person than we possess of her intellectual viewpoint. Such phrases as the following addressed to natural philosophers, which are so characteristic of Cavendish, seem to tell us more about personality than intellectual outlook:

> I wish heartily my Brain had been Richer, to make you, a fine Entertainment . . . and though I cannot Serve you on *Aggat Tables,* and *Persian Carpets,* with *Golden Dishes,* or *Crystal Glasses,* nor Feast you with *Ambrosia,* and *Nectar,* Yet perchance my *Rye Loaf* and new *Butter* may Taste more Savoury, than Those that are Sweet, and Delicious.[2]

While it was of course quite common for early modern women to apologize for the quality of their works, or for writing at all, Cavendish combined her admission of limited abilities with attacks against the intellectual establishment in universities or the Royal Society. It is the seeming transparency of her autobiographical statements, expressed in simple self-revealing language, yet coupled with strong views and sharp criticisms of respected institutions and persons, that most characterize Cavendish. And it continues to be difficult to assess whether the straightforward autobiographical voice or the sharp and iconoclastic assessment of the seventeenth-century intellectual landscape best represents the essential Margaret Cavendish—or whether they are two sides of a quite complicated nature.

Margaret Cavendish has likely received more attention as a scientist than has any other seventeenth-century English woman. And this is the case even though few would term her a scientist, or even a natural philosopher, either because they doubt her intellectual stature altogether, or because they would be more apt first to designate her a poet, a playwright, or an author of fantasies. She has gained most attention from her social status and supposed eccentricities; she has always garnered attention as the first woman to be invited to the Royal Society, and from the heterodox nature of her writings. More recently she has come to be studied as a serious student of scientific and philosophic controversies.[3] But in this essay I want to place her study of science into a context, and conceptual framework, that I have developed elsewhere: namely to link her scientific skepticism to her essential utilitarianism.

This approach emerges from my study of Cavendish's biography, both intellectual and personal. What actually worked for her, and for learning and society more broadly, determined the topics that interested her and the direction she took in studying and making judgments concerning them. Ultimately, her utilitarian bent (even though the term would be anachronistic for her age) meant she had trouble deciphering what seventeenth-century experimental science offered to everyday people, or what it contributed to solving real-life problems. Second, it is important never to underestimate her intellectual questioning of much that she heard or witnessed. The integrated themes, then, that structure this analysis are Cavendish's strong skepticism, which she displayed in her treatment of scientific topics as well as the works of philosophers broadly; her desire for fame, which led her to resist simple acknowledgment of the superior intellectual knowledge or standing in others; and her utilitarian bent, which led her to question both broad philosophical concepts and specific scientific claims.[4]

While Descartes is rightly credited with raising existential doubt to the level of philosophic inquiry, as we know from his *Discourse on Method,* he only pursued such doubt and questioning up to the point that it might force him to question the existence of God. For those who seek a more unrestricted chain of inquiry, Descartes is often criticized for allowing Catholic orthodoxy to truncate his philosophical inquiry. While I would never claim that Cavendish exercised systemic thought comparable to Descartes', still, I would claim that for her little, if anything, was sacred or beyond the bounds of inquiry, and that it was this willingness to question religious, social, and political values that most essentially defined her philosophical stance.

When coming to Cavendish's science from a study of her political writings, as I have done, one can see most clearly the strange and unpredictable views that subverted her predictive class and ideological loyalties. As a duchess, she identified with the poor, with peasants, and with those who would question the right of political philosophers to shape a state based on absolute sovereignty and social hierarchy. As a royalist married to a general in the king's forces during the English Civil War, her arguments in favor of freedom of conscience, and her claim that women had no reason to be loyal to the state, were remarkable. But most remarkable for her intellectual outlook was the following assessment of political philosophers such as Thomas Hobbes, of whom her husband was an important patron:

> . . . when governmental Laws were devised by some Usurping Men, who were the greatest Thieves and Robbers, (for they Robbed the rest of Mankind of their Natural Liberties and Inheritances, which is to be equal Possessors of the World;) these grand and Original Thieves and Robbers, which are call'd Moral Philosophers, or Common-wealth makers, were not only Thieves and Tyrants to the Generality of Mankind, but they were Rebels against Nature, . . .[5]

Such a position was not simply at odds with royalist sentiments, but it follows a sentence so typical for Cavendish, and virtually unique among her contemporaries, namely the use of biblical language to establish that nature, not God, created the universe: "Nature, who made all things in Common, She made not some men to be Rich, and other men Poor, some to Surfeit with overmuch Plenty, and others to be Starved for Want: for when she made the World and the Creatures in it, She did not divide the Earth, nor the rest of the Elements, but gave the use generally amongst them all."[6] Such views were multiply transgressive: they contradicted the biblical story of creation, they spoke to an essential equality reserved to levellers and diggers and their allies on the other side of the revolutionary divide, they questioned the legitimacy of civil society, and they allowed a female nature not simply to create the universe, but to establish its principles. In presenting this position, I need to clarify that I am not arguing that Margaret Cavendish fits within that vision of early modern science most associated with the work of Carolyn Merchant, namely that women were resistant to a rational, linear, empirical vision of knowledge that had made early modern science a masculinist enterprise. She was a woman who resented traditional learning based on scholastic and classical principles, and who ridiculed the scholarship current in universities, but she did not want to replace it with a more ecologically friendly and nurturing variety. She was a strong rationalist who greatly admired philosophers and other scholars, and in her most famous utopian work, *The Blazing World,* while designating an empress as its ruler, she still honored the learned as the empress's most valued advisers.[7]

Most important for her scientific writings were the strong utilitarian strains underlying her assessment of seventeenth-century science, especially experimental science, and her willingness to take on the most renowned thinkers and theories. As I noted earlier, there seemed little sacred to Margaret Cavendish except perhaps her own intellectual standing. As she indicated on innumerable occasions, fame was what she sought and she would use any means to obtain it. This led her to write in a range of genres trying to determine what the public, and especially male intellectual critics, might praise—or accept from a female pen. She adopted different personae, sometimes accepting women's essential inferiority and other times claiming men had always sought to monopolize the tools and institutions of learning, writing her scientific works in poetic form because poetry was more acceptable for a woman, and donating her works to Oxford and Cambridge universities with fulsome praise for scholars resident there. While others have claimed this was due to the contradictions within her own thought, I would argue it was a clever, calculated attempt to gain ultimate acceptance for her work. While it may seem a foolish, and perhaps demeaning, act to donate her works to universities made up of scholars who were likely only to ridicule them, in the long run she had the last laugh because today we have an extant collection of her works not available for other seventeenth-century women. Over the centuries, such donating, as well as her gauging the acceptable, truly were successful strategies to overcome resistance to having herself and her works taken seriously. The complexity and the gendered nature of this approach are found in a poem dedicated to fame that introduces her *Philosophical Fancies:*

To thee, great Fame, I dedicate this Peece.
Though I am no Philosopher of Greece;
Yet do not thou my workes of Thoughts despise,
Because they came not from the Ancient, Wise.

Nor do not think, great Fame, that they had all
The strange Opinions, which we Learning call.
For Nature's Unconfin'd, and gives about
Her severall Fancies, without leave, no doubt.

Shee's infinite, and can no limits take,
But by her Art as good a Brain may make.
Although shee's not so bountifull to me,
Yet pray accept of this Epitome.[8]

By placing nature in God's stead and speaking for her, Cavendish created a forum to legitimate herself, to feminize the infinite, and to use simple but significant markers, such as the comma between "ancient" and "wise," to demonstrate their distinction and the possibility of her superiority to each. And her rejection of orthodoxy is so often tied to gender inversion that it is difficult not to accept that that was one of the reasons behind her particular constructions. For example, as a strong materialist, she posits matter at the core of the universe while others might identify the supernatural. Yet it is a matter, which she often terms "dull matter," that receives its action externally from a number of sources, usually subsumed under the rubric of "motion." And in the midst of a lengthy discussion of the nature of matter and the variety of motions, in a section on "the working of severall motions of nature," she includes the following phrase: "thus may the one change, if Motion hurles That Matter of her waies, for other worlds." She continually engenders powerful and creative elements female, and as the term "hurles" makes clear, her female power within the natural and physical world is not a stereotypically feminine force, but a dominant, commanding authority.[9]

One of the most striking qualities of Cavendish's works are the innumerable prefatory materials that introduced them, filled with apologia or explanations of what readers had failed to grasp in an earlier work; here I would turn only to the preface to *Sociable Letters* as evidence for her shifting personae, and the ways in which she intended such shifts to gain her acceptance and ultimate fame. One can detect how even it defined alterations in her scientific point of view as she later hedges her earlier position about the power and infinity of nature in her *Observations upon Experimental Philosophy*. This essay contends that Cavendish chose the genre of her writings, and the topics on which she wrote, and shifted her point of view, not so much because she was inconsistent or could not make up her mind, but to please her audience. To gain fame, she was willing to listen to critics who said she should avoid science and philosophy, should write in poetry and fantasy, and should highlight God's power and dismiss nature as an independent, infinite force.

Her basic scientific values, or changes in a particular perspective, were often tied to her intellectual reliance on William and Charles Cavendish, the latter her brother-in-law who was interested in mathematics and scientific questions, and most prominently Thomas Hobbes. I am not so sure, however, that her shift away from an early excitement about microscopes, telescopes, and other scientific instruments was based on the influence of Hobbes and others who doubted their worth. Here her justifications are much more utilitarian than substantive or philosophical and reflect a theme that dominated her work more generally.

In turning to the preface to the *Sociable Letters,* one finds Cavendish's fullest efforts to gain acceptance as an intellectual. She begins with a plea to her readers not to follow the pattern of earlier critics of her poetry who "found Fault that the Number was not Just, nor every Line Matched with a Perfect Rime." She explains that there were only a few errors, which "were by Chance," and that in Latin and Greek, poems do not rime but are based "only [on] an Exact number of Feet and Measures." She then picks up a common theme for her, namely that critics of her work have pointed to insignificant points, in this case rimes, which are "as the Garments, and not as the Body of Wit." Others had criticized the placement of words in *The World's Olio,* and she agrees such errors may be common, "for I was not bred in an University, or a Free-School, to Learn the Art of Words." After dismissing such criticism by saying that she will "leave the Formal, or worditive part to fools, and the Material or Sensitive part to wise men," she turns to the criticisms of her *Philosophical Opinions.* Readers complained of its obscurity. First, she states that she does not have the power to bestow wit to enhance their understanding, but she then acknowledges revisions based on such criticism: "I since not only Over-viewed, and Reformed that Book, but made a great addition to it, so that I believe, I have now so clearly declared my sense and Meaning" that only fools could miss them. Others have found fault with her plays because not all of the characters are acquainted. Her *Orations* were criticized for a wholly different and more substantive reason: "for speaking too Freely, and Patronizing Vice too much." But in her defense she claims she is a strong enemy of vice and only wrote as she did to test her fancy, "for surely the Wisest, and eloquentest Orators, have not been ashamed to defend Vices upon such Accounts, and why may not I do the like?"[10]

As these defenses introduce her new work of sociable letters, she anticipates criticism of that work for imitating epistolary works based on "Romancical" forms, but her intent was not to sway audiences with "High Words, and Mystical Expressions," but to provide descriptions of serious issues or real-life events. And, finally, to exemplify again her willingness to shift genre and perspective to gain acceptance, she explains: "But the

Reason why I have set them forth in the Form of Letters, and not Playes, is, first, that I have put forth Twenty Playes already, which number I thought to be sufficient, next, I saw that Variety of Forms did Please the Readers best. . . . And thus I thought this to be the best Way or Form to put this Work into, which if you Approve of, I have my Reward."[11]

Cavendish's hunger for fame, combined with her willingness to shift the nature of her writings to gain acceptance, can offer one model for her shifting scientific views. Her earliest scientific writings were in verse, and some have linked this to a more poetic, or even feminine, nature that distanced her from natural philosophy. But Cavendish offered her own explanation in *Poems and Fancies:* "The Reason why I write it in Verse, is because I thought Errours might better pass there, than in Prose, since Poets write most Fiction, and Fiction is not given for Truth, but Pastime."[12] For those unfamiliar with such verses, they are exemplified in the following titles, and while they may be bad science, they are worse poetry: "Nature calls a Council, which is Motion, Figure, Matter and Life, to advise about making the World," "The four Principal Figured Atomes Make the four Elements," and "Of Contracting and Dilating, whereby Vacuum must needs follow."

In turning to her perspective on the use of the microscope, her utilitarian and skeptical qualities are evident throughout. But they carry distinct import from the earlier period in which she evaluated their worth to her views in the late 1660s as expressed in *Observations upon Experimental Philosophy,* where she found little social or intellectual redeeming values in the microscope. Her utilitarian loyalties led her to question the value of concepts or procedures that did not lead to practical results, improve knowledge of the sensible world, or produce immediate applications. And they led Cavendish to shift her opinion on the usefulness of the microscope. At first she saw it as an instrument with important potential, but in her later writings she was apt to consider it more appropriate to child's play—something with which to offer temporary amusement, and perhaps shock, but not something that could seriously enhance the storehouse of knowledge.[13]

Her interests in natural philosophy and experimental science indeed had been tied to an early desire to understand the operations of the microscope and to associate with scientists who employed it. One of the few extant letters we have of Cavendish is a brief letter to Christiaan Huygens concerning the operation of his glass (microscope). It replies to a second letter she had received from him and reveals both her serious interest in the microscope and the practical problems she sought to answer with its use. It is written from Antwerp and represented her time in exile when both she and her husband had contact with European philosophers and scientists. It is a brief letter, and one difficult to summarize, so it seems wise to reproduce it:

Sir

I have rec d yo r second letter by Mr dewerts wherein I finde yo r dissatisfaction of ye Opinion of those Little glasses; truly Orbs [?]are as obscure and hard to finde out by those that are unlearned in them, as Natures Workes; but to Cleere my Opinion, or rather to Answere yo r desires I shal argue something more of them; though my Arguments may be as weak: as my opinions, & my opinions as weak as my Judgement, & my Judgement as weake as want of Knowlledge Can Make itt. As for the Liquer you say in your Letter that if itt were a sulphorous Liquer, or a Liquid gunpowder (as I said) I thought it might bee, doubtlesse it would be active by which helpe of fire; I answer for that fire hath severall Active Effects both in it selfe and upon other substances; or subick wherfore if ye Liquer had ben drye powder it might be subiect to yt effect of fire, as to flash flame, or burned; but if ye powder were wett, the fire culd work not such effects, but as ye substance is A Liquer fire is as subiect to yt [that] substance, or, matter as that substance or matter is to fire for all Liquer allthough strong wth spiritts and hott in operation will quinch fire as sudenly as fire shall euaperate[,] Liquer take quantity for quantity and it is probably yt the high fire you did Apply in ye glass: , did euaporat out ye Liquid in the glasse which might be ye weakenning & changing or altering the former effects; Also yo u say yo u Cannot perceie ye bubble to be a Liquer, I Answer that it is probable the Licquers, if any of them was Euaperated out either by ye fire, you applyed or by ye vent of passage; which may soone turn itt intoVapor by reason of its Little quantitites yt is in a glasse, thus it might be wasted before ye truth could possible be found out for Certanly to my sense, as also to my reason A Licquer Apeared to be in those glasses; yo u sent me; but if there be noe Licquer in those glasses, then itt is probable it might be pent up ayre enclosed therin, which hauing vent was ye Cause of the sound, or report which those glasses gaue. Thus see [?] you may perceive by my Arguinge, I strive to make my former opinion, or sense good; Allthough I doe not binde my self to opinions, but truth; And ye truth is thus though I cannot find: out of the truth of ye glasses; yet In truth I am.

Antwerp 30 March 1657

Sir

Your humble saruant

M Newcastle[14]

The letter was transcribed by a secretary, and she adds in her own hand that she would have written it herself, "but by reson my hand written is not legabell I though[t] you might rather have gest at what I would say then had read what I had writt." It is interesting, without knowing the experimental context, that she has been taking part in a dispute with Huygens and clearly seems willing to raise doubts concerning microscopic-based research and to offer her explanation, which differs from his more expert and experienced stance. And it is not so clear that his observation and explanation are inherently superior to hers. But, as she moves closer to the publication of her *Observations* in 1666, she is less and less convinced of the microscope's utility. She places her criticism within the context of her husband's skepticism of anything that did not lead to a practical outcome, and her criticism is extreme. While verifying his ownership of microscopic equipment, she continues that he is little involved "with this brittle Art," and that William does not find "the Informations of those Optick glasses . . . so useful as others do."[15] She wonders why men "busie themselves more with other Worlds, then with this they live in." It is a strange enterprise unless they find some means "that would carry them into those Celestial Worlds, which I doubt will never be." She concludes in this introductory portion of her work that "I have but little faith in such Arts, and as little in telescopical, Microscopical, and the like inspections, and prefer rational and judicious Observations before deluding Glasses and Experiments."[16]

In the section of her work "Of Micrography, and of Magnifying and Multiplying Glasses," Cavendish outlines in greater detail her distrust of experimental science. After saying that she does not use such instruments, she continues that this art, "with all its Instruments, is not able to discover the interior natural motions of any part or creature of Nature; nay, the questions [*sic*] is, whether it can represent yet the exterior shapes and motions so exactly, as naturally they are; for Art doth more easily alter then inform."[17] She continues with the construction of cylinders and concave and convex glasses which represent objects not "exactly and truly, but very deformed and mishaped." Such instruments reproduce "hermaphroditical" phenomena "as partly Artificial, and partly Natural." For an example she offers: "a Lowse by the help of a Magnifying-glass, appears like a Lobster, where the Microscope enlarging and magnifying each part of it, makes them bigger and rounder then naturally they are."[18] But her more important criticism, rather than simple misrepresentation, is the issue of what is to be gained from such an enterprise.

Most importantly, microscopes only enhance our knowledge of the surface of an animal or object, and they distract from more serious learning. Their use "has intoxicated so many mens brains" and led them to dwell on "but superficial wonders." Her basic utilitarian skepticism follows:

> But could Experimental Philosophers find out more beneficial Arts . . . either For the better increase of Vegetables and brute Animals to nourish our bodies, Or better and commodious contrivances in the Art of Architecture to build us Houses, or for the advancing of trade and traffick to provide necessaries for us to live . . . [they] would not onely be worth their labour, but of as much praise as could be given to them.[19]

But these instruments have not demonstrated such practical import, and rather the scientists who employ them could be described:

> But as Boys that play with watry Bubbles, or fling Dust into each others Eyes, or make a Hobby-horse of Snow, are worthy of reproof rather than praise, For wasting their time with useless sports; so those that addict themselves to Unprofitable Arts, spend more time then they reap benefit thereby.

She suggests instead that they take up husbandry, architecture, "or the like necessary and profitable employments."

The major failing of microscopes was their inability to discover the interior structure of the observed object. "[A]s for the interior form and motions of a Creature, . . . they can no more represent them, then Telescopes can the interior essence and nature of the Sun."[20] No one, for instance, can determine the constituent parts of milk by viewing a drop underneath a microscope. It is only through furthering our understanding of an object's true nature that scientific instruments can further knowledge. She concludes this section by returning to the centrality of nature and the inability (and impracticality) of scientific instruments to affect or improve upon her creation.

> Wherefore the best optick is a perfect natural Eye, and a regular sensitive perception, and the best judge is Reason, and the best study is Rational Contemplation joyned with the observations of regular sense, but not deluding Arts; for Art is not onely gross in comparison to Nature, but, for the most part, deformed and defective, and at best produces mixt or hermaphroditical figures, that is, a third figure between Nature and Art: which proves, that natural Reason is above artifical Sense, as I may call it.[21]

Cavendish's utilitarian skepticism with microscopic research is tied to her claim to intellectual independence. She knew that women were excluded from both philosophic training and academic circles, and she used that exclusion as a forum to differ with men who possessed such training. Her search for fame was tied both to resentment of, and pride in, her ability as a woman working in her isolated study to read and assess natural philosophers and experimentalists. She consistently displayed a wariness of those deemed expert in an area of study. Thus Margaret Cavendish exemplified an almost unique position among women intellectuals in that she admired (or at least would like to take the place of) university dons and those acknowledged as philosophers and scientists, but at the same time she roundly criticized them for relying too heavily on the ancients (in the one instance) and on inadequate, unnatural instruments (in the other).

Her desire for fame demanded that she claim the originality of her works, so she denied taking her thoughts from others, especially Hobbes and Descartes, about whom such charges had been made. Much of her writings reflect a kind of common sense tied to a general knowledge of natural philosophy. She applied broad principles of the character of nature, of matter, of motion, of atoms, and the like, to specific descriptions of animals and physical phenomena. Reading Cavendish's scientific writings often loses one in the trees and obscures the forest, especially when trying to track consistencies through numerous discussions of matter or motion in a range of works written in different genres. The two significant tropes that characterize her throughout are her common sense/utilitarian perspective and her willingness to comment on virtually any phenomenon or set of beliefs.

Cavendish takes on a range of values, especially religious, without care as to whom she might offend. And this combination of utilitarianism and open questioning often leads her to views beyond the intellectually or socially acceptable. For instance, in comparing her with Lady Anne Conway (usually identified as the most important late seventeenth-century English female metaphysician), Conway is clearly the better educated, and likely the more intelligent of the two, but the range of her questioning was limited by her adoption of Christian cabbalism. She knew classical languages and authors, as well as contemporary philosophers, but her questioning is circumscribed by her continual reference to cabbalist sources, as well as her dominant goal to understand God's creation and his glory. In her *Principles of the most Ancient and Modern Philosophy, concerning God, Christ, and the Creatures . . . of Spirit and Matter in General,* which was published posthumously in 1690, she begins with the chapter "Concerning GOD, and his Attributes," which focuses on traditional qualities of an omnipotent God. She states:

> GOD is a Spirit, Light and Life, infinitely Wise, Good, Just, Mighty, Omniscient, Omnipresent, Omnipotent, Creator and Maker of all things visible and invisible.[22]

Conway's work, which is devoted to a lengthy description of the nature of God's creation and the material and spiritual nature of earthly creatures, continually references book one of the Kabbala. It has a number of interesting points, especially its sense of the basic unity and spirituality of all created beings. However, it does not move beyond a fulsome discussion of God's special attributes and the necessary creation that emerges from them. It is because "God is infinitely Good, Loving and Bountiful" that his creatures carry a strong spiritual nature and that there is a basic unity among all within that creation. Such views allow her to situate a strong spirituality among animals and to discover links among the lowest and highest of God's creation. Her analysis emerges from the principle that "It is an Essential Attribute of God to be a Creator." Yet while incorporating a greater erudition and making more systematic arguments, her work neither questions the basic values of Judeo-Christianity nor raises the gender issues present throughout Cavendish's work. It is also ironic given the limited numbers of women able to read Latin that the translator of her work into Latin in 1690 (that preceded a 1692 English version) noted that his translation of the work meant "that thereby the whole World might be in some sort benefitted."[23]

As Cavendish begins her *Philosophical Letters,* written in 1664, she moves away from her earlier claims for nature, but never adopts the language of a Judeo-Christian creation. She continues to portray a powerful nature, but one that is subordinate to a more powerful God—but in just what ways superior she is somewhat vague. While nature is infinite, she is not supreme or composed of the highest qualities. Cavendish explains:

> God . . . has not onely Natures Infinite Wisdom and Power, but besides, a Supernatural and Incomprehensible Infinite Wisdom and Power."[24]

Neither interferes with the function of the other. In an early section of the *Philosophical Letters,* she takes some pains to show God's greater power: "For I do verily believe, that there can be but one Omnipotent God, and he cannot admit of addition, or diminution; and that which is Material cannot be Immaterial, and what is Immaterial cannot become Material, I mean, so, as to change their natures; for Nature is what God was pleased she should be; and will be what she was, until God be pleased to make her otherwise."[25] But the extent of her movement away from an infinitely powerful nature to an even more powerful infinite deity is not totally clear as she continues to employ phrases such as "Infinite Matter, which has Motions and Figures inseparably united."[26]

Here it is doubtful as to how thoroughly she undercuts her earlier viewpoint. Clearly her friends, and perhaps indirectly her critics, told her that she had gone too far in making nature the center of her universe and that she needed to stipulate God's role in creation. Yet she continues to term nature infinite and to separate her from God and not treat her as a mere creation of God. At the least there seems some confusion in her mind as to the independent qualities of God and of nature, and a strong sense that she resisted giving up this powerful being, gendered female. Even in supporting God's superiority, she justifies such a decision upon the authority of divinity identified as "she." Cavendish sought her own fame, and she believed it could only come through the possibility of female power and authority, and this made it difficult for her to develop a view of creation, or of knowledge, emerging from a patriarchal God. The reach of that ambition is made clear in her famous comment to her readers in *The Blazing World:*

> For I am not covetous, but as Ambitious as ever any of my Sex was, is, or can be; which makes, that though I cannot be Henry the Fifth, or Charles the Second, yet I endeavour to be Margaret the First; and although I have neither power, time nor occasion to conquer the world as Alexander and Caesar did; yet rather then not to be Mistress of one, since Fortune and the Fates would give me none, I have made a World of my own: for which no body, I hope, will blame me, since it is in every ones power to do the like.[27]

It is thus necessary to integrate an understanding of her personal and intellectual goals, her realization of the fact that as a woman she could not attain serious intellectual recognition, and the way in which both her unrestricted questioning and her desire for fame influenced her work to begin to grasp the unorthodoxy and complexity of her scientific writings and their troubled reception during the 1600s and today.

## Notes

1. There are a number of works that outline Margaret Cavendish's life and intellectual interests and works. They include the traditional biography by Douglas Grant, *Margaret the First: A Biography of Margaret Cavendish, Duchess of Newcastle, 1623–1673* (London: Rupert Hart-Davis, 1957), and more recent biographies and analyses of her intellectual efforts: Anna Battigelli, *Margaret Cavendish and the Exiles of the Mind* (Lexington: University Press of Kentucky, 1998); and Katie Whitaker, *Mad Madge: the Extraordinary Life of Margaret Cavendish, Duchess of Newcastle, the First Woman to Live by Her Pen* (New York: Basic Books, 2002). However, even with these recent studies, mostly because of the lack of her personal correspondence, we have limited understanding of the nature of her life and her intellectual motivations.

2. Margaret Cavendish, Duchess of Newcastle, "To Natural Philosophers," *Poems and Fancies, The Second Impression, much altered and corrected* (London: printed by William Wilson, Anno Do. MDCLXIV [1664]).

3. A recent treatment of her scientific thinking appeared in Whitaker's chapter on "Queen of Philosophers, 1667–1673," 303–46, and the section on her importance as a natural philosopher in the collection *A Princely Brave Woman: Essays on Margaret Cavendish, Duchess of Newcastle,* ed. Stephen Clucas, a scholar who has long recognized her scholarly contribution during the 1600s (Aldershot, UK: Ashgate, 2002), 185–254.

4. Earlier elements of this analysis, especially queries concerning her autobiographical writings and the degree to which they should be taken at face value, have appeared in Hilda L. Smith, "'A General War amongst the Men [but] None amongst the Women': Political Differences between Margaret and William Cavendish," in *Politics and the Political Imagination in Later Stuart Britain: Essays Presented to Lois Green Schwoerer,* ed. Howard Nenner (Rochester: University of Rochester Press, 1997), 143–60; and in Smith, "The Management of the Grange, the Household and Medicine: Common Sense and the Fantastical Margaret Cavendish" (paper offered at the second international Cavendish Symposium, University of Paris, June, 1999).

5. Margaret Cavendish, *Orations of Divers Sorts, Accomodated to Divers Places* (London, 1662), 85–88.

6. Ibid.

7. The most recent edition, along with a useful introduction and commentary, is *The Description of a New World Called the Blazing World and Other Writings,* ed. Kate Lilley (London: William Pickering, 1992).

8. Margaret Cavendish, "A Dedication to Fame," *Philosophical Fancies* (London: printed by Tho: Roycroft, for J. Martin and J. Allestrye, 1653).

9. Ibid., 29–30.

10. Margaret Cavendish, "The Preface," *CCXI Sociable Letters* (London: William Wilson, 1664).

11. Ibid.

12. Cavendish, "To Natural Philosophers," *Poems and Fancies.*

13. For discussions of her shifting scientific perspectives, see the introduction to *Observations upon Experimental Philosophy,* ed. Eileen O'Neill (Cambridge: Cambridge University Press, 2001), x–xxxvii; and Stephen Clucas, "Variation, Irregularity and

Probabilism: Margaret Cavendish and Natural Philosophy as Rhetoric," in Clucas, *A Princely Brave Woman,* 199–209. Clucas's work emphasizes the similarity of Cavendish's thought to many of the thought and stylistic patterns of contemporary male thinkers with similar interests, and especially her solid place within seventeenth-century attitudes concerning probability.

14. British Library, Add. Ms. 28558, f. 65.

15. Margaret Cavendish, "To His Grace the Duke of Newcastle," *Observations upon Experimental Philosophy. To which is added, the Description of a New Blazing World.* (London: A. Maxwell, 1666).

16. Ibid.

17. Cavendish, *Observations upon Experimental Philosophy,* 7.

18. Ibid., 8.

19. Ibid., 10.

20. Ibid., 11–12.

21. Ibid., 12–13. Here, she is clearly in agreement with the perspective of Thomas Hobbes in his dispute with the accuracy and utility of Robert Boyle's air pump as argued most thoroughly in Steven Shapin and Simon Schaffer, *Leviathan and the Air-Pump, Hobbes, Boyle and the Experimental Life* (Princeton: Princeton University Press, 1985).

22. Anne Conway, *Principles of the most Ancient and Modern Philosophy, concerning God, Christ, and the Creatures . . . of Spirit and Matter in General* ([Latin] Amsterdam, 1690; repr., [English] London, 1692).

23. Ibid., 11–12. For an insightful and sympathetic analysis of Conway's thought, see Sarah Hutton's essay comparing her with Margaret Cavendish in her coedited (with Lynette Hunter) collection, *Women, Science and Medicine, 1500–1700: Mothers and Sisters of the Royal Society* (Stroud: Sutton, 1997), 218–34.

24. Margaret Cavendish, *Philosophical Letters, or, Modest Reflections upon some Opinions in Natural Philosophy* (London, 1664), 9.

25. Ibid., 10.

26. Ibid., 11.

27. Margaret Cavendish, "To the Reader," *The Description of a New World, called the Blazing World* (London: printed by A. Maxwell, 1666).

JUDITH P. ZINSSER

# The Many Representations of the Marquise Du Châtelet

In the seventeenth century the "querelles des femmes [argument about women]" generated spirited advocacy for women's intellectual capacities. Learned men, such as François Poullain de la Barre, published tracts and essays to support their unorthodox views. In "De l'égalité des sexes" (On the Equality of the Sexes), he insisted that women had the capacity to reason and learn. Only their inferior education kept them subordinate to men. Despite such arguments, the centuries-old negative and denigrating gender assumptions whether ecclesiastical, legal, or popular held sway.[1] Little had changed in the next hundred years. With numerous texts and traditions to draw from, even in the elite world of Europe's eighteenth-century "Republic of Letters" the *philosophe,* the *géomètre,* and the *physicien* were male by definition. "*Les gens savants* [the learned men]" of France's Académie royale des sciences (Royal Academy of Sciences) did not expect a female to exhibit the intellectual qualities of reason, observation, and reflection, nor to acquire the skills in mathematics, astronomy, and mechanics that could warrant her inclusion among their ranks. An eighteenth-century French woman, however privileged and learned, would have to make decisions about how to represent herself should she ever dare to publish on the subject of "natural philosophy." Contemporaries would readily criticize her ideas, dismiss her conclusions, and ridicule her pretensions, simply because she was female. Gabrielle Emilie le Tonnelier de Breteuil, marquise Du Châtelet (1706–49), chose to challenge these demeaning presumptions and negative expectations. With her first extended work, the *Institutions de physique,* published in 1740, she demonstrated not only her mastery of scientific and philosophic books and treatises but also a range of intellectual abilities traditionally reserved to men.[2]

How did she accomplish this? A combination of circumstances made possible the serious consideration of her ideas even though put forward by a woman. Her brilliant use of conventional methods of validation achieved the rest. In the *Institutions de physique,* Du Châtelet presented images of an author well versed in the key works of natural philosophy, able to manipulate the formulas of advanced mathematics, and familiar with experimental physics. Thus, she gained praise from her reviewers and acceptance as a participant in the philosophical and scientific debates of her day. It was the *Institutions* and Du Châtelet's subsequent victories in the public challenges to her ideas that caused future writers, such as the mathematician Jean le Rond d'Alembert and the philosopher Emmanuel Kant, to mention her specifically when presenting their own views on *forces vives* (kinetic energy).[3]

## The Malleability of "Science"

Du Châtelet's goal in the *Institutions* was to present a coherent, and thus unitary, complete system of the universe, a description that challenged narrow, mechanistic explanations of nature's laws. Indeed, she wanted to explain everything: from the smallest bit of matter to the substance and behavior of the planets and their satellites, from how we know what we know to God's role in every aspect of the universe. To do this she drew on her wide reading in what today would be categorized as physics, mathematics, chemistry, philosophy, and metaphysics. She believed that all were necessary to explain how we perceive and explain the workings of the cosmos. Like many of her contemporaries she saw no need to separate these kinds of inquiry. Quite the reverse. As she explained in her preface: "Many truths of Physics, Metaphysics, & Geometry are evidently interlaced *[sont evidemment lieés entre elles]*."[4]

Du Châtelet purposely embraced the metaphysical tradition of Descartes, Leibniz, and Christian Wolff. She borrowed freely from their ideas of the bases of knowledge and of God's role in the universe, from their methods of reasoning, and from their concepts of the essential nature of matter, space, and time. She described Galileo's and Newton's mechanics, the astronomical observations of Kepler, and the work of French mathematicians such as Pierre Louis Moreau de Maupertuis. With this original synthesis of what appeared to many contemporaries as apparently contradictory methods and concepts, she hoped to bridge what she and many Continental critics of Newton's *Principia (Mathematical Principles of Natural Philosophy)* perceived to be the enormous gap in his published writings between his mathematics and mechanics and a fully explained system of the world. For her and so many others, a study of the cosmos and its governing laws required more than description, for example, an explanation of its originality. Why this and only this universe? Why a universe that functioned according to these laws and no others?

Although it would not occur to most of us to consider these kinds of questions when thinking about our universe from a scientific perspective, Du Châtelet lived at a time when "science" as we understand it had not yet developed. "Les grands philosophes," as Du Châtelet called them—thinkers such as Newton and Leibniz—tried to reason out a unified system.[5] Others, like the Dutchmen Wilhem Musschenbroek and Hermann Boerhaave, shunned such efforts to explain the unverifiable questions of origins and first causes as occult tendencies reminiscent of the thirteenth-century Scholastics. Instead they mixed substances, measured phenomena, recorded their observations of the material world, and refused to speculate further. Voltaire, Du Châtelet's companion and lover during these years, favored this latter group,[6] and Du Châtelet, the former. However, a simple description of these broad differences in approach can be misleading. In the 1730s and 1740s in France, the learned took ideas from many sources and many disciplines. *Philosophe* (philosopher), *géomètre* (mathematician), and *physicien* (physicist), all had broad meanings not necessarily synonymous with what is meant by those words today, or even what they meant a decade later in 1750 or at the end of the eighteenth century.

Mathematics, chemistry, astronomy, physics, and mechanics, in that era, were viewed as new empirical categories of inquiry. The *Journal des sçavans* listed only "philosophia" and "scientiae" in its index of 1741; by 1759, philosophy, mathematics, and astronomy had become separate entries, and natural history had been added. Books and journals of the first half of the eighteenth century laced their titles with a variety of descriptive phrases. Terms such as "experimental" and "natural" philosophy were only the most common. Others appear to have been synonymous. For example, successive manuscript versions of Du Châtelet's *Institutions* show that she changed "physique" to "science." Similarly, for chapter headings in her commentary on the *Principia,* either she or her editor changed "philosophie" to "physique," despite Newton's preference for the former term.

To further complicate our understanding of their thinking, learned men distrusted "experiment," or *"expérience,"* as it is called in French. To Du Châtelet and the authorities she read, although empirical knowledge was essential, the number of possible variables could defeat even the most determined experimental efforts. The senses could receive biased impressions. The reliability of results depended on the qualifications of the demonstrator and the observers of the *expérience.*[7] Descartes, Christiaan Huygens, Leibniz, and subsequent thinkers like Du Châtelet relied on the reasoned and abstract language of mathematics, which quantified experiments and created "axiomatic truths," a necessity before accepting the simplest conclusion from observation. Neither pure deduction nor pure induction, as Du Châtelet argued in the *Institutions,* could alone contribute to the progress of *"les sciences."*[8]

At its inception and in the subsequent reform at the end of the seventeenth century, the Académie royale des sciences had been encouraged by royal ministers to take a narrow, pragmatic view of science. The King favored the prospect of practical applications of the Académie's members' work. Therefore, statements of purpose in its resolutions and publications remained adamant in the exclusion of all philosophy, theology, and metaphysics.[9] Despite the King's preference and these clear pronouncements, however—and the Académie's strict categorization of the disciplines into three mathematical sciences (geometry, astronomy, and mechanics) and three experimental sciences (anatomy, chemistry, and botany)—the members tolerated extreme differences of opinion (Cartesians and Newtonians, for example), and many continued to explore subjects beyond what could be observed or quantified.[10] Maupertuis read papers on specific problems in astronomy and mathematics at the Académie meetings—held on Wednesdays and Saturdays—but continued to think and to write in more speculative and analytical ways. All kinds of writings continued to be published, some even as *mémoires,* the official documents of the Académie. For example, the three prizewinning essays on fire from the biennial competition of 1738 presented very different views of its nature and differed in turn from Du Châtelet's in the essay she had submitted, and hers from Voltaire's entry.

This malleability of the term "science" and of the range of views that were tolerated by most and accepted by many had intellectual consequences that worked to the advantage of a woman. It made validation possible for someone like Du Châtelet, whose synthetic approach and sources of authority crossed so many cultural and disciplinary boundaries, drawn from a combination of German, French, Italian, Dutch, and English metaphysics, astronomy, chemistry, mechanics, and mathematics.[11] However, Du Châtelet went beyond demonstrating familiarity with these authors and their works. Instead she integrated their methods, concepts, and discoveries into her text and used them to illustrate the underlying premises of her own natural philosophy. For example, she used Kepler and Newton to demonstrate a rhetorical point about how we reason: how hypothesis worked in discovering the orbit of the planets. A circular path had been hypothesized, but observations turned into mathematical calculations proved it contradictory, thus necessitating the search for a new hypothesis and its subsequent confirmation. Similarly, she described how Galileo's experiments with falling bodies led to his formula for their rate of descent, a theory "at present adopted by all Philosophers, and for which each experiment has become a demonstration."[12]

## The Persona of the "Man of Science"

Circumstances favored Du Châtelet's acceptance in other ways. For not only was the definition of "science" malleable, so also was the persona of the "man of science."[13] Most of the new generation of Academicians had of necessity learned on their own, or from mentors. Few attended universities or joined their faculties. The intellectual profile of these new members of the Académie in the 1730s was of a young man, about twenty-eight years old, with perhaps a few years of formal advanced schooling (at a local *collège,* the prelude to university training), who had acquired most of his education from tutors, and then as the protege of a member of one of France's twelve provincial academies or the Académie in Paris. In many ways, this describes Du Châtelet's situation.[14]

Thus, much of the new French experimental and mathematical science in these decades developed outside of traditional, formal institutions.[15] There was, in fact, a network of the learned, from London to Berlin to Basel to Bologna to St. Petersburg and the other capitals of Europe. Fathers and sons like the Eulers and brothers like the Bernoullis made up a large extended family engaged in puzzling out the most perplexing aspects of the universe. Members of this network validated each other and each other's conclusions.[16] When asked by their colleagues, they wrote the "Approbation" needed for the approval of France's royal censors before publication. In the same way the mathematician Henri Pitot supplied the "Approbation" for Du Châtelet's *Institutions,* and Alexis Clairaut for her *Principia* translation and commentary. They reviewed each other's books in the myriad learned journals of the day, just as they would review Du Châtelet's. Such attention gave credence and authority to their work, so it did to hers.[17]

Du Châtelet had cajoled and forced her way into this network. Adopting the practices of her male contemporaries, she corresponded widely with the authors of the books and treatises she read. She requested their works, asked their opinion of ideas she favored, made clear her disagreements, footnoted them in her writings, and sent them her own published books and essays. She invited them to visit and to tutor her. With Voltaire's financial support and her husband's approval, she created her own Académie at Cirey, the dower estate that became a refuge for them from the fall of 1735, off and on until 1740. Here, she entertained the learned men of science, *"les Emiliens,"* as Voltaire called them. Maupertuis and Alexis Clairaut, two of the most prominent of France's mathematicians; Francesco Algarotti, whose *Newtonianismo per le dame (Newton's Philosophy for the Ladies)* inspired Voltaire's *Eléments de la philosophie de Newton;* Johann Bernoulli, whose theories Du Châtelet presented in her *Institutions;* and Père Jacquier, a coeditor of the Geneva edition of the *Principia* (1739–42), were only some of her most important visitors.

This description is not meant to contradict historians' depictions of the masculine traditions of Europe's intellectual history, and the subsequent gendering of learning; rather, it suggests that for a brief period those traditions and that gendering could be made less consequential. For a little over a decade, a woman, the marquise Du Châtelet, participated in and was recognized for her activities within the masculine world of the *"divers gens sçavans* [diverse learned men]" and was acknowledged as the possessor of intellectual qualities usually identified as male.

As a whole, the *Institutions de physique* reads like a virtuoso performance in which Du Châtelet presents herself as *philosophe, géomètre,* and *physicien.* On page after page she demonstrated her knowledge of the major philosophical and scientific works familiar to those trying to puzzle out a coherent "system of the world." Aside from the obvious references to the authorities that she considered "*les grandes hommes* [the great men]," Descartes, Leibniz, Wolff, and Newton, at different points she also cited Locke *(Essay on Human Understanding)* and Samuel Clarke (his published interchange with Leibniz); made analogies to the history of Alexander the Great and the Romans; referred to Plato, Aristotle, Zeno, Diogenes, Virgil, and St. Augustine; used many of Nicolaas Hartsoecker's observations with the microscope and experiments of early chemists such as Robert Boyle and René-Antoine Réaumur; agreed with Democritus, Epicurus, and Gassendi on particles and with Archimedes on mechanics; and explained the discoveries of famous astronomers including Ptolemy, Copernicus, Galileo, Kepler, Cassini, Huygens, James Bradley, and their commentators like Jacques Rohault and John Keill. Citations throughout the *Institutions* showed her familiarity with the *mémoires* of the French Académie, the *Philosophical Transactions* of the British Royal Society, Germany's *Acta Eruditorum,* and the papers of the Russian Academy of Sciences in St. Petersburg—and with the assertions of men forgotten by all but specialized historians of science such as Varignon, Picard, Richer, Malezieu, Raphson, Freind, Fermat, Wallace, Blonde, Norwood, and Hermann. Each of these references, presented as they were as part of her grand synthesis, demonstrated the unity of her approach and gave authority to her explanations and conclusions.

## Authority by Association: The *Philosophe*

In particular the first ten of the twenty-one chapters of the *Institutions* display Du Châtelet's abilities as a *philosophe.*[18] Here she maintained the unity between metaphysics and science. As her explanations in the *Institutions* unfold, she showed that she derived her method from a surprising range of apparently contradictory sources. The majesty of these chapters, concerning what we would now categorize as philosophy and metaphysics, lies in her synthesis of authorities her contemporaries saw as in opposition, of their methods of reasoning, and of their first premises.[19] Often this meant that she accepted one aspect of an author's ideas but rejected others.

For example, although initially identified with French supporters of Newton, Du Châtelet explicitly disagreed with his ideas in a number of instances. As mentioned previously, she embraced the importance of hypothesis. Though modern scientists would now approve of her insistence on its efficacy, many European followers of Newton and Descartes railed against what they considered to be the last vestige of "Scholasticism," the tool of superstition and error. Her definition and advocacy of "hypothesis" was both a measure of her ability to take authority and a sign of her understanding of

how science had evolved and would continue to evolve. Hypothesis, she believed, was "the thread" that leads to the "purest" discoveries. Although she acknowledged that "the good hypotheses are then always the work of great men," such as Copernicus, Kepler, Huygens, Descartes, Leibniz, and Newton, the efforts of others also had their value. "It is necessary," she explained, "that some among us risk being led astray, in order to mark a good path for others." She recognized the dangers: one successful experiment could not confirm an hypothesis, but one that did not support the initial supposition would be enough to justify its rejection.[20]

Perhaps most interesting was her use of Descartes. Many of her contemporaries saw only disagreement between his writings and those of later natural philosophers such as Newton. She incorporated ideas from both. Although she used an imperious tone when dismissing Descartes' second law of motion—"this rule is contradicted by experiment"—and later would dismiss his theory of particles and "tourbillons"—flowing currents of matter—as the motive force of the planets, from Descartes she took her belief in reasoning from first principles.[21] Like him, she accepted that a sound use of one's reason, or *"l'entendement,"* as she called it, could alone constitute justification and proof. She explained to her readers that these first principles came largely from Leibniz and Wolff, but she reasoned much as Descartes recommended—from the simplest to the more complex, very much like a geometric proof. She would have agreed with historian of philosophy Desmond Clarke's characterization of this approach: that "metaphysics as a theory of knowledge establishes the possibility of scientific knowledge both in physics and mathematics," that it is only by understanding first causes that one can proceed to secondary consequences and abstract descriptions of their reality.[22] In this, she clearly allied herself with those of her contemporaries who favored understanding not only how the universe worked but also why it worked as it did.

## Authority through Mathematics: The *Géomètre*

Du Châtelet was probably introduced to Euclidian geometry at the age of ten or eleven, sitting in with her brother's lessons. Then as a young married woman in 1732–33, like others of her circle, Du Châtelet took lessons in advanced mathematics from the charming *géomètre* Maupertuis. Unlike her friends, however, she embraced the study with enthusiasm and went on to learn calculus. When Maupertuis was unavailable, she convinced Clairaut to be her tutor as well. At one point in February of 1739, deep within the extensive revisions to the early chapters of the *Institutions,* stymied by the alternative explanations of motion in the universe, Du Châtelet wrote to Frederick of Prussia that she was returning to the study of mathematics, for "geometry is the key to all the doors & I am going to work to acquire skill in its use." She had already written in January to Maupertuis that after his having "given [her] the extreme desire to apply herself to geometry and to calcu-

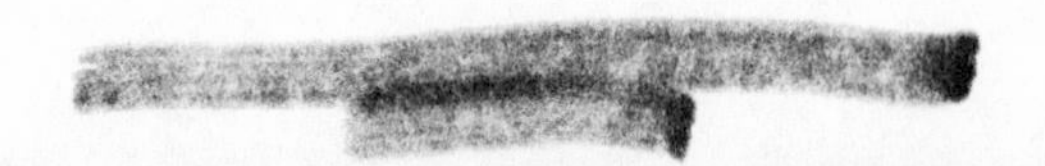

lus," she wished for him to encourage his Swiss colleague, Johann Bernoulli, to come to Cirey to instruct her, "to bring light to bear on her shadows."[23] However, when Bernoulli did not join her household, she was undeterred and continued on her own.

As a *géomètre,* she believed that the use of mathematical abstractions eliminated the risks inherent in natural philosophy. To this end, she embraced Newton's original use of analytic geometry, what I. Bernard Cohen, the modern translator of the *Principia,* calls "the Newtonian Style." Du Châtelet, like Newton, saw that this geometry provided the means to move from the concrete experiences of the experimenter to the abstract hypotheses of the speculator, and thus to create mathematical idealizations of the real world. Proportions and ratios drawn from Euclidian and analytic geometry described particular relationships which became general principles that could be presumed to apply in all instances. It was the means, she explained, by which our minds gain the power to make abstractions. She neatly integrated Newton's geometric explanations from the *Principia* throughout the second half of her text.[24] One could begin with an observation, an experiment, a question. Like many of Newton's commentators, Du Châtelet used his example of the man walking along a road to initiate a discussion of the relationship between time, speed, and distance. Geometry and mathematics in general could be used in discovering not only the basic laws of mechanics but also those governing the planets and their satellites. For example, she noted, it was "with the aid of the most pure geometry" that Newton explained the success of Kepler's Second Law.[25]

The geometric formula or geometric figure, according to Du Châtelet, could always make the particular truth "easier to see."[26] For her, mathematics was the equivalent to fact, certain knowledge on which subsequent hypotheses could be founded. Thus, she, along with Leibniz, believed that it could be used for metaphysics as well as for mechanics. Where metaphysicians of the past would have used scripture, she used Euclidian geometry. In this sense geometry replaced scripture; its laws could be applied to all reasoning and thus lead to more certain explanations and experiments. She thought it apt that Plato called the Creator *"l'éternel Géomètre* [the eternal mathematician]."[27] She understood that Leibniz had been seeking "a calculus for Metaphysical truths, by means of which, solely by the substitution of characters, one could arrive at these truths as in Algebra." Perhaps she imagined herself contributing to his project.[28]

Geometrical figures and relationships provided her with the unimpeachable analogy that proved metaphysical explanations. For example, in Chapter 1 on "The Principles of Our Knowledge," both the concept of the "law of sufficient reason" and "of continuity" had their geometric analogy. The second is the most elaborate. To prove Leibniz's assertion that an entity passing from one state to another must go through all the gradations of change, Du Châtelet described a line shifting from concave to convex, then determinants of an ellipse approaching those of a parabola.[29] Elsewhere, qualities of essences are analogous to the nature of quadrilaterals; "extension" is a line that becomes geometric surfaces, "cube ABCDEFGH"; the simple division of a line into its constituent points demonstrates the divisibility of matter.[30]

Finally, she seems to have taken her style and method of argument from geometry as taught to her by her second mathematics tutor, Clairaut. We know of Clairaut's lessons from the geometry text he published long after their lessons together had ceased.[31] In his preface he explained that one should proceed step by step, describing how an explanation had developed, and allow "*les Commenças* [the beginners]" to follow a "route similar to those of the inventors." As he explained, "Following this path, beginners perceive at each step that one makes them take, the Inventor's reason, & by this means they can more easily acquire the spirit of invention."[32] Similarly, in the *Institutions* Du Châtelet presented the historical evolution of ideas, for example, on "matter." In this way readers could follow in the way she had been led in her study of geometry.[33] In addition, her numbered paragraphs and sections, like a geometry text, moved from one axiom or corollary to another. Just as in geometry, she established one of her premises and then built upon it to establish the next and the next. Thus, Du Châtelet presented herself as her own authority and provided self-constructed validation for her explanations of phenomena.

## Authority from Experiment: The *Physicien*

Clairaut also insisted in his text that the best way to understand phenomena was to go back and forth between abstraction and what could be perceived with the senses. For example, he began with a practical exercise, how surveyors measure the land, which in turn introduced the student to abstraction in the form of the "principal truths of geometry."[34] In the *Institutions,* Du Châtelet adapted this technique to her own purposes. She moved back and forth between abstraction and its opposite, the evidence of the senses, to explain and to prove the laws of physics governing the natural world. Thus, she proved herself a *physicien,* capable of making clear and compelling arguments not only with mathematics and hypothesis but also from observation and *expérience.*

Du Châtelet, like other natural philosophers, took from Galileo both the language of observation and the faith in experiment as confirmation of the hypotheses that resulted. Throughout the *Institutions,* she referred to experiments, both those of others, like Huygens's work with the pendulum, and her own.[35] Her experiments on the nature of fire account for many of her references: to the properties of mercury and steel; to the heating and cooling of sulphur, silver, and gold; to the effects of the "*machine de vuide,*" or air pump. One can almost picture her in her room with various instruments and objects around her. The pendulum clock on her chimney breast perhaps facilitated her many references to clockworks.[36]

Was it a vase sitting on a nearby commode filled with flowers from her gardens at Cirey that contributed to her demonstration of the figuration of bodies? She described all of the manifestations of water, from a substance in a glass to a vapor, to snow, to ice, and again to the liquid poured into just such a vase.[37]

Du Châtelet also explained how it was possible to move from fabricated experimental situations to analogies with the real world, and thus to formulation of new truths about the universe. She did this with her presentation of Galileo's ideas, and more extensively with Newton and his application of the forces of terrestrial dynamics to the universe as a whole. For example, she described the sequence from Newton's experiments with gravity, to an hypothesis about the shape of the earth, to its relationship to the observations of the astronomers Picard and Richer, to the final corroboration for his supposition that attraction flattens the earth at the poles with Maupertuis's expedition to Lapland.[38]

But even as Du Châtelet lauded such experiential corroboration, she cautioned her readers about the potential hazards of using sensory observation and experiment to validate one's suppositions. They could lead to error as easily as truth. Only results systematically arrived at demonstrated the correlation between hypothesis, mathematical abstractions, and reality. "It is," she explained, "indispensably necessary in order to preserve oneself from error to verify one's ideas, to demonstrate them as reality, & not to admit as certain what one would not have assured oneself by experiment or by demonstration had been affirmed as neither false nor chimerical."[39] For example, our senses might deceive us, as when a moving object on the shore seems to be still as we move at the same speed: "their images always occupy the same points on our retina."[40] Results of experiments might seem random and inexplicable. In seeking the reason for a particular phenomenon, one should expect to correct the possible explanation many times. Like Descartes, she wondered if perhaps "the secrets of Nature will elude us" altogether because of the multiplicity of possibilities.[41] Nevertheless, she encouraged her readers: "the efforts that one makes to find the truth are always glorious, even if they could be without fruit."[42]

In addition to references to formal experiments, Du Châtelet drew on other kinds of experience for corroboration—informal, commonsense observations. These analogies have a particular vividness in the text.[43] To understand the difference in perception of the part versus the whole, she offered the example of the chorus in the opera that gives more pleasure to those in the loges than those in the "*coulisses* [wings]"—if one is so close, "one hears particular voices and loses the combination of voices *[l'ensemble]*" that makes the beauty.[44] To explain "*lieu absolut* [absolute place]" versus "*lieu relatif* [relative place]," she offers analogies from her immediate surroundings. The paper on which she writes exemplifies how we might have "a clear idea" but must be open to the discovery of those aspects hidden and contained within that idea, "for there are an infinity of things in this, the texture of this paper." A table, a bed, chairs exist, but one cannot always tell which changed place when there is a new arrangement. In the same way, she continued, we cannot decide if the sun turns around the earth or the earth around the sun, because the appearance is the same for both suppositions.[45] Though modern science now has different explanations for these phenomena, perhaps it was a late night of writing that suggested the following analogies to her. Attraction explains the movement of the planets, so, she continued, "By the same principal, oil rises in the cotton of a lamp, ink adheres to my pen, and sap rises in plants."[46]

Du Châtelet also referred to her own body, if the analogy seemed apt. The concept of "mode" in Chapter 3 first led her to the clock face and then to her arms. She *could* move them anywhere in theory, but, in fact, their actual movements depend on exterior objects and on the position they were in just before. The way in which "*nôtre Ame* [our soul]" receives impressions she likened to the partial, and therefore obscure and confused, impressions a child receives in the uterus, where the slightest movement of the mother's body is unexplained and undifferentiated.[47]

Even the Leibnizian metaphysical concept of "sufficient reason," the concept that for her was the "touchstone which distinguished truth from error," Du Châtelet made seem self-evident with an analogy to her own situation and her own reasoning. She explained that no one decides for one thing over another without a sufficient reason that makes him see that this thing is preferable to the other. To deny this principle, she insisted, meant that we fall into "*absurdités* [absurdities]." There would be no reason for anything to last for more than an instant. To her it seemed obvious. "For example," she writes, "I can be lying down, standing, it must be that there is a sufficient reason why I am standing & not sitting or lying down."[48]

## Authority from Controversy

That Du Châtelet cited authors of varying views in her synthesis did not mean that she avoided the controversies over "the most important truths of Physics & Metaphysics" known to the natural philosophers from whom she sought validation.[49] Instead, she drew authority from making her disagreement explicit, even when the author was one she favored in another chapter on a different subject. Aristotle she identified as "the father of Philosophy and of error," certainly "a great man, but he was a bad Physicist."[50] While an advocate of Locke's views on perception, she thought his ideas on "Substance" "entirely confused," and opposed his potentially heretical suggestion that "*matière* [material objects]" might have been given the attribute of thought, extended by some of his readers to mean that all matter had a soul.[51] Unlike Voltaire and many Newtonians in England and France, she insisted that undemonstrable causal questions about the workings of the universe could be resolved.[52] Unlike many members of the Académie, she rejected Descartes' "tourbillons" (vortices carrying the planets in their orbits), but like Descartes, she stopped short of claiming "certain" knowledge of the workings of nature "true" or "certain" in the same sense as known to God. Here, she likened our inability to understand divine purpose to the way in which the eye works: it "cannot see the smallest particles of an object without losing the view of the whole."[53]

By her discussion of these disagreements, Du Châtelet not only demonstrated her knowledge and her ability to reason but also adroitly placed herself in relation to the men she cited and their controversies. If she accepted some of their ideas and rejected others, she showed herself to be above faction, independent of those who took one position or another. Thus, by this rhetorical device she not only presented herself as the neutral commentator but also established herself as the ultimate authority. Throughout, her tone and phrasing set the relationship. M. Mariotte was simply wrong in the eleventh Proposition of the second part of his Treatise on Percussion. She then proceeded to give the correct analysis. She complimented M. de Crousaz, winner of an Académie prize for his essay on time, for "many excellent things," but then "cannot imagine how he could have said" what he did. She instructed us, and by implication M. de Crousaz, to examine ideas "with the eyes of Understanding," and to be wary of "this confusion which gives birth to almost all the objects that are evident to our senses, & whose realities are often infinitely different than appearances."[54]

As in the case of Crousaz and others, especially in the chapters on mechanics, she took authority by explaining how others came to their erroneous observations or conclusions. This ability to explain their errors automatically set her yet higher above those she corrected. M. de Frenicle was misled, his results entirely false, because "a badly conceived experiment threw him into error." "This example," she continued, "makes us see that we must be even more circumspect about the experiments that we do, because our conceit always speaks to us in our favor." A wise caution for the *physicien,* but humiliating to the recipient nonetheless.[55]

The final chapter of the *Institutions,* in which Du Châtelet gave a detailed description of the arguments over *forces vives,* shows her willingness to take on the great controversies of her day and her consummate skill in denigrating the opposition and in placing herself in the role of the wiser, more accomplished *géomètre* and *physicien.* In Chapter 21 on "The Force of Bodies," she chose to disagree with no less an authority than Jean-Jacques Dortous de Mairan, the man who succeeded Bernard le Bovier de Fontenelle as "perpetual secretary" of the Académie royale des sciences. Her attack on him is unique: she devoted the entire chapter to demolishing his treatise on *forces vives,* an Académie *Mémoire* of 1728. Unlike previous controversial topics, she offered no explanation for her opponent's errors and abandoned any semblance of neutrality. Instead she was an obvious defender of a particular faction. Her tone was both adversarial and authoritative, admitting no room for doubt; "reason shows us and experience confirms" her, not his, conclusions.[56] The view she supported, that of Johann Bernoulli, whose treatises also appeared in Académie publications (1724, 1726), was presented as a truth, even though some continued to disagree with it. His view, she explained, was "the sole point of Physics," still disputed despite the experiments that prove its validity. By implication, only a few misguided individuals, for example Dortous de Mairan, insisted on supporting what she called *"absurdités."* She explained that she had chosen his treatise to answer in detail because he argued with "much clarity *[clarté]* and eloquence," and because it was "the most ingenious against *forces vives.*" Mairan had, she told her readers, "said all that one could say in favor of a bad cause."[57]

She quoted long passages from his treatise to discredit both his arguments and his abilities as a scientist. She mocked him with his admission that he "is not at all afraid to advance [this hypothesis], a thing entirely impossible." In contrast, the explanation she favored was very simple and easily perceptible *[trés sensible]*."[58] His reasoning betrayed him, she concluded, and it was, she acknowledged, "seductive," but it flew in the face of a whole host of significant demonstrations. She wanted her readers to understand that Dortous de Mairan's insistence on the false reading of experiments and thus his false conclusions left him isolated from all of the other, and as she implied, better, *physiciens,* not only in France but in all of Europe.[59]

## Du Châtelet's Temporary Victory and Her Significance to the History of Philosophy and Science

The first review of the *Institutions de physique* appeared in the *Journal des sçavans* in two parts. The length alone suggested acknowledgement of the seriousness of the work. Part I, published in December of 1740, made the fact that Du Châtelet was a woman a positive. Her *Institutions* united the "*Esprit Philosophiques* [Philosophic Mind]" usually reserved to men with the graces usually given only to women. The reviewer described her as a "genius," not a "prodigy," a significant distinction in the eighteenth century, when the former usually described a man and the latter a woman, and thus presented an anomaly.[60] Part II of the review appeared three months later in March 1741. The author more or less ceased editorial comments and instead continued to go through chapter by chapter, quoting at length. The reviewer acknowledged the fact of disputes among *"d'illustres Sectateurs* [illustrious Followers of a sect]" as illustrated by Du Châtelet's text, but refused to favor one view or the other. Instead it was left to the reader to judge the merits of the arguments.

In retrospect, the passage of time appears significant because in the intervening months the Académie had published a pamphlet by Dortous de Mairan answering Du Châtelet's attack on his views on *forces vives.* As she noted in a letter to a friend, comte d'Argental, the attack, however denigrating its words and patronizing its tone, was in itself a compliment.[61] By implication it suggested that her writings had become too authoritative and too significant to be ignored. Her challenge to Mairan's views of *forces vives* in her essay on fire and then at length in the *Institutions de physique* seems to have been too much for the eminent *physicien* to overlook. Newly appointed to his post at the Académie, he probably assumed that he could easily dispense with this female challenge to his and the prevailing views of motion. Thus, he could maintain his intellectual reputation and his status among the learned elite.

The *Mémoires de Trevoux* reviewed the *Institutions* in May of 1741, ignored the dispute, and did not even mention that *"L'Auteur Anonyme"* was a woman.[62] Rather, the *"éxtrait"* began, and continued throughout, to applaud the style and thus the usefulness of the book: "rendered with much finesse of style, & *feu d'imagination* [imaginative ardor], two rare qualities, but always estimable in books made to be read for the general Public." It complimented the author's presentation of *"le systême"* of Leibniz, a subject "little known and not very fashionable in France," in "an instructional, colloquial tone, at ease, intelligible & meanwhile noble and full of propriety *[bienséance]*."[63] Chapter by chapter, this "good Compilation of modern Physics," is "wise & discreet," "full of intelligence & refinement of the mind *[finesse d'esprit]*," in that it does not go to extremes, even when defending Johann Bernoulli's position in the *force vives* argument, "so to speak, with vivacity, against the most learned adversaries."[64]

In a subsequent review of Du Châtelet's published interchange with Mairan, however, the *Mémoires de Trevoux* first complimented her ability to answer him point for point, "reason for reason, witticism for witticism," and to use irony with such mastery. Then the journal's reviewer took a deprecating tone, quoted her examples out of context, listed the learned opponents of her views, and indicated that her train of argument was very good, by implication, for a woman.[65] Despite the denigrating nuances of this reviewer, and the echoes of traditional attitudes towards women's intellectual abilities, the *Institutions* is often credited with having introduced Leibniz's ideas to French audiences and formed the basis of Du Châtelet's election to the Bologna Academy of Sciences. Her translation of and her commentary on Newton's *Principia* (published posthumously in 1759) gained similar praise. The *Journal Encyclopédique* called it "dazzling, contributing to the moment of triumph of the philosophy that it explains and comments on."[66]

And Du Châtelet's significance to our understanding of the gendering of knowledge and of learned institutions? For our histories of science in the age of the Enlightenment? Carolyn Merchant Iltis made a convincing case for Du Châtelet's specific contribution to the resolution of the *forces vives* controversy. Recently, popular historian of science David Bodanis suggested that she had almost formulated the law for the conservation of mass decades before Lavoisier.[67] But the writings and career of the marquise Du Châtelet have a broader significance. She is part of the missing chapters of the traditional history of science, the chapters that explain the transformation of natural philosophy into its modern counterparts of metaphysics, philosophy, and science. Her writings and their positive reception are yet more proof of the continuing role of theological and philosophical speculations in the scientific thought of the first decades of the eighteenth century. As scholar of the social history of science Simon Schaffer has insisted, the development of modern science in France did not simply involve Cartesians and Newtonians. Whether identified as *philosophes, physiciens,* or *géomètres,* natural philosophers like Du Châtelet were "representative of a distinct philosophy" in its own right.[68]

The marquise Du Châtelet is part of another narrative as well. Her career exemplifies the missing chapter of learned women's active participation in the first decades of the Enlightenment, not only as facilitators of men's intellectual endeavors, but also as agents and contributors in their own right. This chapter of European intellectual history ends not only with the transformations and separations of knowledge into new disciplines, the creation of new institutions and new definitions of the learned, but also with the exclusion of women. Older traditions about women's incapacities prevailed and once again denied even the elite access to training and refused to validate their discoveries or to engage with their arguments. Though a few women continued to think and work in these fields, they gained attention from the learned only within a carefully circumscribed and gendered world that placed them and their activities second to those of men. Past women of achievement like Du Châtelet fell away from common memory as historians of science and philosophy created seamless linear narratives that left room neither for female genius nor for the mixing of mechanics and metaphysics.

## Notes

1. On Poulain de la Barre, see Siep Stuurman, *François Poulain de la Barre and the Invention of Modern Equality* (Cambridge, MA: Harvard University Press, 2004). For a detailed description of the denigrating gendered tradition, see Ian McLean, *The Renaissance Notion of Woman: A Study of the Fortunes of Scholasticism and Medical Science in European Intellectual Life* (New York: Cambridge University Press, 1980); and Lieselotte Steinbrügge, *The Moral Sex: Woman's Nature in the French Enlightenment,* trans. Pamela E. Selwyn (New York: Oxford University Press, 1995). For general descriptions of the phenomenon of women's exclusion, see Mary Terrall, "Gendered Spaces, Gendered Audiences: Inside and Outside the Paris Academy of Sciences," *Configurations* 2 (1994): 207–32; and Londa L. Schiebinger, *The Mind Has No Sex? Women in the Origins of Modern Science* (Cambridge, MA: Harvard University Press, 1989).

2. Contemporaries first learned of her scientific expertise in her long review letter to the *Journal des sçavans* in September 1738. The letter, in theory, was to bring Voltaire's *Eléments de la philosophie de Newton* to the attention of the elite, but she also cited key writers and controversies, mentioned problems still to be solved, and offered her own ideas about solutions. *Journal des sçavans,* September 1738 (Arsenal 4o H.8909), 534–41.

In 1738 and again in 1740, the Académie published the essays that she and her companion, Voltaire, had submitted on the nature and propagation of fire, for its biennial prize. Their essays appeared as fourth and fifth, after the three winners, Leonhard

Euler, Luzeran du Fiesc, and the comte de Crequy. *Recueil des pieces qui ont remporté le prix de l'Académie depuis 1720* (Paris: chez Claude Jombert, 1732–77), vol. 4. On reception of this and other publications, see Keiko Kawashima, "Madame Du Châtelet dans le journalisme," *LLULL* 18 (1995): 471–91. I am grateful to Roger Hahn and Natalie Zemon Davis for making me aware of this article.

3. Mary Ellen Waithe, "Gabrielle Emilie le Tonnelier de Breteuil du Châtelet-Lomont," in *A History of Women Philosophers,* vol. 3, *Modern Women Philosophers, 1600–1900,* ed. Mary Ellen Waithe (Boston: Kluwer Academic Publishers, 1991), 127–51. See especially 142–43.

4. [Marquise Du Châtelet], *Institutions de physique* (Paris: Prault fils, 1740), 14; see also Chapter 7, "On Matter," 148. Scholar of comparative literature Erica Harth describes her approach in this way: "In metaphysics she replaces the Cartesian idea of God as first causal principle by Leibnizian sufficient reason, while in physics she replaces Descartes by Newtonian mechanical principles and Leibnizian dynamics." Erica Harth, *Cartesian Women: Versions and Subversions of Rational Discourse in the Old Regime* (Ithaca: Cornell University Press, 1992), 197. However, this summation is misleading. Du Châtelet made "God" very much part of her "system of the world." An omniscient God played the key role in her universe, not *"l'artisan,"* tending the springs and cogs of a machine, but *"l'ouvrier,"* the wisest and most perfect creator whose works mirrored his qualities and therefore needed no adjustment. See *Institutions,* 175, 45–47, 52. The whole of Chapter 2 concerns her views of God, his qualities, and his role. See on this question more generally Mary Terrall, "Metaphysics, Mathematics, and the Gendering of Science in Eighteenth-Century France," in *The Sciences in Enlightenment Europe,* ed. William Clark, Jan Golinski, and Simon Schaffer (Chicago: University of Chicago Press, 1999): 246–71; and Wolfgang Lefèvre's introduction to *Between Leibniz, Newton and Kant: Philosophy and Science in the Eighteenth Century* (Boston: Kluwer Academic Publishers, 2001), vii–xvi.

5. See the description of Newton's efforts in Betty Jo Teeter Dobbs, *The Janus Face of Genius: The Role of Alchemy in Newton's Thought* (New York: Cambridge University Press, 1991). On this aspect of Du Châtelet's intellectual development, through reconstruction of the successive revisions of the *Institutions* from 1738–40, see Linda Gardiner [Janik], "Searching for the Metaphysics of Science: the Structure and Composition of madame Du Châtelet's *Institutions de physique, 1737–1740," Studies on Voltaire and the Eighteenth Century* 201 (1982): 85–113.

6. The first edition of Voltaire's glib, engaging *Eléments de la philosophie de Newton* (1738) exemplified this approach. For the complete version, including his *Traité de Métaphysique,* see the 1741 critical edition, edited by Robert L. Walters and W. H. Barber, in *The Complete Works of Voltaire* (Oxford, UK: Voltaire Foundation, 1992), vol. 15. The *Eléments* was a collaborative effort between Voltaire and Du Châtelet. She is assumed to have been largely responsible for helping him to understand the mathematics of Newton's works. They disagreed on key aspects of the theories he presented, and these disagreements in all likelihood prompted Du Châtelet to write and publish her own book about the workings of the universe.

7. Steven Shapin and Simon Schaffer, *Leviathan and the Air-Pump: Hobbes, Boyle, and the Experimental Life* (Princeton, NJ: Princeton University Press, 1985), 333–36. On Descartes and experiment, see Desmond M. Clarke, *Descartes' Philosophy of Science* (Manchester, UK: Manchester University Press, 1982), 71, 140, 20–21. Simon Schaffer, "Self-Evidence," in *Questions of Evidence: Proof, Practice, and Persuasion across the Disciplines,* ed. James Chandler, Arnold I. Davidson, and Harry Harootunian (Chicago: University of Chicago Press, 1994), 56–91. See also David Gooding, Trevor Pinch, and Simon Schaffer, eds., *The Uses of Experiment: Studies in the Natural Sciences* (New York: Cambridge University Press, 1993).

8. See *Institutions,* 106. Also drawn from Simon Schaffer, "Natural Philosophy" in *The Ferment of Knowledge: Studies in the Historiography of Eighteenth-Century Science,* ed.

G. S. Rousseau and Roy Porter (New York: Cambridge University Press, 1980), 85, 88–89; Harold J. Cook, "The new philosophy in the Low Countries," in *The Scientific Revolution in National Context,* ed. Roy Porter and Mikulás Teich (New York: Cambridge University Press, 1992), 136; and J. L. Heilbron, *Elements of Early Modern Physics* (Berkeley: University of California Press, 1982), 21.

9. See Roger Hahn, *The Anatomy of a Scientific Institution: The Paris Academy of Sciences 1666–1803* (Berkeley: University of California Press, 1971), 16, 50.

10. Hahn, 98–99. Note that Du Châtelet's frontispiece for the *Institutions* showed allegorical figures for the same six sciences.

11. A number of scholars have commented on Du Châtelet's method. The most recent is Julie Hayes, who explains that "she appropriates or puts to use texts, without possessing them" (108). This comment suggests that Hayes, as a literary critic, may not have perceived the nature of Du Châtelet's original synthesis. See Julie Candler Hayes, *Reading the Enlightenment: System and Subversion* (New York: Cambridge University Press, 1999). Carolyn Merchant Iltis saw it as Newtonian "mechanical principles, Leibnizian dynamics and a combination of Descartes, Leibniz and Newton's natural philosophy" (31). See Carolyn Merchant Iltis, "Madame du Châtelet's Metaphysics and Mechanics," *Studies in the History and Philosophy of Science* 8 (1977): 28–48. On the significance of method as a means of validation in this transitional/malleable period, see Ludmilla Jordanova, *Languages of Nature: Critical Essays in Science and Literature* (London: Free Association Books, 1986), 30.

12. See *Institutions,* 75–79, 89, 83, 84, 259–60. Contemporaries would have enjoyed the irony of the example from Newton, given his often quoted rejection of "hypothesis" in the General Scholium of Book 3 of the *Principia.*

13. The Académie des sciences was not the closed world suggested by its historical reputation or that some of its members hoped to create. It was only during the first decades of the eighteenth century, under the direction of its "perpetual secretary," Bernard Bovier de Fontenelle, that the modern image of the "scientist" was crafted. His "*Eloges* [Eulogies]" of deceased members gradually constructed our more modern definition of the "the man of science." See Charles B. Paul, *Science and Immortality: The Eloges of the Paris Academy of Sciences (1699–1791)* (Berkeley: University of California Press, 1980). See also Mary Terrall, *The Man Who Flattened the Earth: Maupertuis and the Sciences in the Enlightenment* (Chicago: University of Chicago Press, 2002), especially chapter 1, for a general description of how Maupertuis fashioned an identity for himself as a "man of science."

14. See Paul, 75; 76, table 3; 97n40. In contrast, by 1780, the average age would be about thirty-nine and the profile similar to what we would expect in the twentieth century with most education in a formal academic setting. See also David J. Sturdy, *Science and Social Status: The Members of the Académie des Sciences 1666–1750* (Woodbridge, UK: Boydell Press, 1995). A number of physicists in France and England gave public series of lectures on Cartesian and Newtonian science, e.g., abbé Nollet in Paris. Du Châtelet certainly knew Nollet and purchased instruments from him; she may have attended some of these lectures.

15. On this phenomenon reaching back into the seventeenth century, with men such as Pascal, Descartes, and Fermat, see L. W. B. Brockliss, "The Scientific Revolution in France," in Porter and Teich, 61, 69.

16. An essay would be identified as having been "read in the Assemblies," listened to by "diverse Learned Men" (Hahn, 62).

17. On the power of writing to establish credibility, see Roger Chartier, ed., *A History of Private Life: Passions of the Renaissance* (Cambridge, MA: Harvard University Press, 1989), 382.

18. Du Châtelet positioned herself from the very beginning of the *Institutions.* She started modestly in the Preface as the anonymous author, a parent writing for a son. In a few pages, however, she established herself as outside and above the contro-

versies of her day, and thus able to judge and to decide between all of the theories offered to date. See, for example, Preface, 4–5, and "On Monades," 131. Hayes gives an excellent analysis of Du Chatelet's textual devices and "model for knowledge," as she positions herself as one of the learned elite: 87–88, 108–9.

19. In the Preface she cautioned her son that authorities in and of themselves are not infallible. Both Aristotle and Newton had surprised her with their *"absurdités."* "The use of reason," she explained, meant that "one must always examine for oneself and set aside the esteem *[la considération]* that the opinion of a famous man always carries." *Institutions,* 11.

20. See *Institutions,* 84, 89. Note that all of Chapter 4 is "On Hypothesis."

21. See *Institutions,* 36–37.

22. Clarke, 102; see also 81, 172.

23. Du Châtelet to Frederick, 27 February [1739], D1912, *The Complete Works of Voltaire,* ed. Theodore Besterman (Oxford, UK: Voltaire Foundation, 1970), 90:257; Du Châtelet to Maupertuis, 20 December [January] 1739, D1804, *Complete Works of Voltaire,* 90:105.

24. See Chapter 5, "On Space," *Institutions,* and for specific examples, 180, 409. Note that she mastered the *Principia* so well that she ignored Newton's order of argument and explanation. Instead, she used the complicated Latin text to suit her own organization. His Book 1 appears in her Chapters 11, 12, and 13. His Book 2 appears throughout Chapters 11–19. His Book 3 is incorporated in pieces throughout: Newton's Rules of Reasoning are in Chapter 1; the description of the workings of the universe in Chapters 11–13, and particularly in Chapter 16. On the "Newtonian Style [Method]," with examples of how it worked in specific sections of the *Principia,* see I. Bernard Cohen, "A Guide to Newton's *Principia,*" in *The Principia: Mathematical Principles of Natural Philosophy,* trans. I. Bernard Cohen and Anne Whitman, assisted by Julia Budenz (Berkeley: University of California Press, 1999), 302–50; and George E. Smith, "The Newtonian Style in Book II of the *Principia,*" in *Isaac Newton's Natural Philosophy,* ed. Jed Z. Buchwald and I. Bernard Cohen (Cambridge, MA: MIT Press, 2001): 249–98, especially 251.

25. *Institutions,* 292.

26. *Institutions,* 262.

27. *Institutions,* 33; see also 142. At this time Du Châtelet was also working on her extensive, chapter by chapter critique of the Bible. Her negative view of the God of the Old Testament and her skepticism about the divinity of Jesus may explain her enthusiasm for mathematics as a model of the universe.

28. *Institutions,* 151. She seems to have believed that Leibniz had developed such a system but had died before revealing it. Clairaut in his geometry textbook expressed a similar goal. Alexis Clairaut, *Elemens de géometrie* (Paris: David fils, 1741), xiii. On Leibniz's joining of metaphysics and scientific observation, see for example Stuart Brown, "Leibniz: Modern, Scholastic, or Renaissance Philosopher?" in *The Rise of Modern Philosophy: The Tension between the New and Traditional Philosophies from Machiavelli to Leibniz,* ed. Tom Sorell (Oxford, UK: Clarendon Press, 1993): 213–30.

29. *Institutions,* 32–35; 24–25.

30. *Institutions,* 179–80.

31. The most obvious instance of echoes from their lessons is her use of Clairaut's example of a man in a boat watching the shore to demonstrate the phenomenon of relative speed. A colleague, Anna Klowoska, drew my attention to Descartes' use of this example as well. See *Institutions,* 109–10, 217; and Clairaut, 2–3.

32. Clairaut, vi–vii, viii–ix.

33. See her Chapter 7, "On Matter," *Institutions,* 129–31.

34. Clairaut, xii. Algarotti's *Il Newtonianismo per le dame* (1736) used these kinds of everyday examples, as had its precursor, Fontenelle's *Entretiens sur la pluralité des mondes* (1686).

35. It is important to note that she rarely used one type of argument in isolation; rather she neatly interlaced different kinds of proof to describe and validate a particular scientific explanation. See Chapter 18, "On the Oscillation of Pendulums," *Institutions,* especially, 375–79.

36. See *Institutions,* 56, 108. Clockworks were a common analogy in the eighteenth century. Leibniz used them, and one of Du Châtelet's most vivid uses of the image is in her description of his concept of "Monades," the essential indivisible bases of all matter. See also 134–35.

37. *Institutions,* 204.

38. Galileo is discussed in Chapter 13, "On Gravity," and Newton in Chapter 15, "On the Discoveries of M. Newton on Gravity," *Institutions.*

39. *Institutions,* 20.

40. *Institutions,* 219.

41. *Institutions,* 196.

42. *Institutions,* 83.

43. Mary Hesse discusses the use of metaphor and analogy in scientific writing and its significance when literal descriptions are inadequate. See chapter 4 of Mary Hesse, *Revisions and Reconstructions in the Philosophy of Science* (Bloomington: Indiana University Press, 1980).

44. *Institutions,* 167.

45. *Institutions,* 108–9.

46. *Institutions,* 146, 323.

47. *Institutions,* 64.

48. *Institutions,* 197, 22, 25. Note that Leibnizian "sufficient reason" explained for her the reason for this universe, these laws, and their predictability.

49. See, for example, her description of the argument over the nature of "space": Was it a void, or filled with some kind of "subtle matter"? *Institutions,* 90–91, 92–93, 108.

50. See *Institutions,* 257–58, 288.

51. *Institutions,* 71, 66.

52. See, for example, Chapter 8, "On the Nature of Bodies," *Institutions,* 178.

53. *Institutions,* 50; on Descartes, see Clarke, 138–39.

54. *Institutions:* for Mariotte, see 268–69, 283; for Crousaz, 125.

55. *Institutions:* on observation generally, see 149; for Frenicle, 285–86.

56. See Chapter 8, "On the Nature of Bodies," *Institutions,* 157. See Hayes's excellent reading of Du Châtelet's rhetorical devices, 91, 104–8.

57. *Institutions,* 428, 426. Dortous de Mairan's essays were known for their clear syntheses of others' views on a particular scientific subject, and thus were often more widely read than the denser, more original *mémoires.* This may have influenced Du Châtelet's choice of his essay.

58. *Institutions,* 433, 431, 434. Note that in this chapter she also argued against the English mathematician James Jurin, a supporter of Mairan's position; see 442 and 442–44 for full discussion. See also her subsequent letter to Jurin, Cirey, 30 May 1744, D2982, *Complete Works of Voltaire,* 93:115–20.

59. *Institutions,* especially 431, 433, 444. In this chapter, she also criticized Newton for his acceptance of the idea that there is a gradual loss of force in the universe. In a neat turn of phrase, Du Châtelet called this "superstition" (a charge often leveled against Newton by those who opposed Leibniz), because of the perpetual intervention by the Creator that Newton's theory necessitated, a system of "continual miracles." See 445–46, 448–49.

60. "Extrait: Part I," *Journal des sçavans* (Paris: Chez Chaubert, 1740), 737.

61. Du Châtelet to d'Argental, 22 March [1741], D2450, *Complete Works of Voltaire,* 91:451.

62. In the first edition of the *Institutions* with the author identified as a parent writing for a son, readers could easily presume that a father, not a mother, was speaking. In the reprint and second editions, Du Châtelet's name appeared on the title page.

63. "Extrait," *Mémoires de Trevoux* (May 1741), 894, 895, 895.

64. "Extrait," *Mémories de Trevoux* (May 1741), 905, 909, 926.

65. The reviewer explained that her intelligence, spirit, and style all should make it impossible for her opponent to ignore that her response "[offers] only the very serious, the very knowledgeable, & very probable for the Learned, among whose ranks madame la Marquise du Châtelet [sic] wants to be and among whom she wants the honor of being counted henceforth." He complimented her on the use of "cette petite figure ingénieuse que nous nommerions volontiers *Figure Françoise . . . Ironie." Mémoires de Trevoux* (August 1741), 1390–1402. See also Kawashima, 472. On the circumstances of the interchange and other scholarly comment, including praise for her book from Clairaut and Claude Helvétius, see Elisabeth Badinter, *Les Passions intellectuelles: desirs de gloire (1735–1751)* (Paris: Fayard, 1999), 1:167–81.

66. *Journal Encyclopédique* 6, partie 3 (September 1959) (Arsenal 8oH2620), 4.

67. Merchant Iltis argues that metaphysical context such as Du Châtelet's made it possible for later eighteenth-century thinkers to move to the concept of two kinds of force, not one (46–48). David Bodanis, *E=mc2: A Biography of the World's Most Famous Equation* (New York: Walker & Company, 2000), 61, 63–67. In this way Merchant Iltis and Bodanis disagree with Voltaire scholar William H. Barber, who, while documenting the significance of Du Châtelet's introduction of Leibniz to French readers, described her as "a disciple" rather than a contributor. William H. Barber, "Mme du Chatelet and Leibnizianism: The Genesis of the *Institutions de physique,"* in *The Age of the Enlightenment: Studies Presented to Theodore Besterman,* ed. W. H. Barber, J. H. Brumfitt, R. A. Leigh, R. Shackleton, and S. S. B. Taylor (London: Oliver and Boyd, 1967).

68. Schaffer, "Natural Philosophy," 71; see also 68–69, 70. Much of my thinking on "originality" and thus on the evaluation of Du Châtelet's contributions comes from the article by Berenice A. Carroll, in which she categorizes the criticisms of women and the lack of praise for the "combinational." See Berenice A. Carroll, "The Politics of 'Originality': Women and the Class System of the Intellect," *Journal of Women's History* 2 (1990), especially, 144, 146–47.

# SECTION II:

# SHIFTING LANGUAGE, SHIFTING ROLES

MARGARET J. OSLER

# The Gender of Nature and the Nature of Gender in Early Modern Natural Philosophy

The gender of nature became the subject of scholarly attention in the early 1980s when some historians and philosophers of science undertook to articulate a new, feminist interpretation of science based on certain aspects of the history of science.[1] These scholars quite naturally turned their attention to the period of the Scientific Revolution (roughly 1500–1700) because, according to the prevailing historiography, that period was a major turning point in the history of science, the terminus ad quem towards which ancient and medieval science led and the termus a quo from which modern science developed. Some feminists argued that a worldview especially repressive to women developed along with the foundations of modern science. In particular, they pointed to the purported change from an organic philosophy of nature to one based on mechanistic principles. They emphasized the rhetoric associated with both the new mechanical philosophy and the Baconian call for an experimental method as particularly denigrating to feminine principles in the abstract and to women in the concrete.

Recent scholarship has demonstrated that this period of the history of science was far more complicated than historians had previously thought. While the older historiography emphasized the internal development of scientific ideas, newer studies emphasize the role of context—intellectual, social, and political. While the traditional interpretation claimed that the Scientific Revolution essentially involved the development of demarcations between science and religion and between the rational and the mystical or occult, recent scholarship has established deep interconnections between theology and natural philosophy as well as an abiding importance of alchemy and other so-called occult studies in the thinking of many of the major figures of this period.[2] These developments have led historians to question the historiographical framework that provided the underpinning for some of the pioneering feminist analyses. To criticize these studies in light of later scholarship is not to dismiss feminist analysis per se. Rather, it is a call for studies based on a revised understanding of what early modern natural philosophy was all about.

Although the early feminist studies do not offer identical analyses of the role gender and metaphors of the gender of nature played during the period known as the Scientific Revolution, they have introduced certain ideas that have become commonplace and are frequently repeated as authoritatively established by other writers who cite them uncritically.[3] Among the propositions that have achieved the status of received wisdom are the following: that nature is portrayed as a woman whom the experimental scientific tradition—in an effort to force her to reveal her secrets—has treated as an apt subject for torture;[4] that the mechanical philosophy of the seventeenth century was particularly negative, destroying the feminine soul of nature and giving rise to the exploitation of the natural world as well as underwriting the decline of the status of women;[5] and that organic philosophies of nature—particularly alchemy—that were supplanted by mechanism had portrayed the feminine in a more positive way.[6] These propositions are often cited as reasons why women have either been excluded from the scientific enterprise or have found it difficult to find a comfortable home within it. In what follows, I will examine the merit of each of these claims.

## Metaphors of Nature and the Gender of "Nature"

Central to the argument about the gender of nature in early modern natural philosophy is the claim that the metaphors used to describe nature carried social baggage that had mostly negative implications for attitudes towards women.[7] A central characteristic of these metaphors—the characteristic that is taken to identify them with the social condition of women—is the feminine gender of the nouns *"terra"* and more generally *"natura."* Associating the feminine gender of these Latin nouns with questions about the status of women, Carolyn Merchant and others have argued that the historical roots of scientific thought have been marked by attitudes demeaning to women.[8]

Does the gender of the noun *"natura"* have any special significance? More generally, what does it mean for nouns to be gendered, and does their gendering have any particular significance?

The practice of categorizing nouns by gender was not related to differences between the biological sexes or the social expectations accruing to them. Rather, early linguists categorized nouns on the basis of their structure. Thus, Latin nouns ending in "a" in the nominative singular follow a particular pattern and are lumped together in a single declension. The nouns that have this form and take adjectives of a certain form are called feminine because the word for "woman," *"femina,"* shares that form as well. But this does not mean that all the feminine nouns carry the cultural baggage about masculinity and femininity that goes along with *"femina."*[9] A question worth pursuing is whether speakers of languages that have gendered nouns ascribe social and psychological meanings to those genders.

A number of the nouns particularly pertinent to early modern natural philosophy are first-declension feminine nouns. Examples include: *"philosophia," "scientia," "physica," "ethica," "logica," "geometria," "mathematica," "regula,"* and *"patria."* Other nouns that have some significance for these questions—*"potestas," "vis," "ratio," "ars,"* and *"lex"*—are also feminine, although they belong to different declensions.[10]

But why is it that *"natura"* as well as many of the nouns referring to the study of nature are feminine? Considering the term *"scientia,"* Londa Schiebinger has suggested that

> Scientia, then, is feminine in early modern culture because it is feminine in the language, but also because the scientists—the framers of the scheme—are male: the feminine *Scientia* plays opposite the male scientist. In order to unite in creative union with the female, the male scientist images his science as his opposite. But more than that, the scientist imagines that a feminine science leads him to the secrets of nature or the rational soul.[11]

Schiebinger finds the sources of the male's search for his opposite in Renaissance Christian Neoplatonism, which depicted the soul as feminine, "the life force of the physical body, whether of the man or of the world. The marriage of the soul to God brought harmony to the universe; the marriage of the philosopher to *Scientia* brought knowledge."[12] The creative force, for these Neoplatonists, results from the union of the masculine and feminine principles. Thus the male researcher sought union with the female *Scientia.* This imagery influenced later natural philosophers even if they no longer accepted a Neoplatonic worldview. Schiebinger illustrates the feminine portrayal of science with an abundance of frontispieces in which "Science" or one of the specific sciences like astronomy or chemistry is portrayed as a woman. Although these examples seem to provide powerful evidence for her claim, she herself notes that there are many exceptions to the feminine portrayal of Science.

There are several problems with Schiebinger's analysis. The term *"scientia"* has a long history, extending back to classical Latin. Defined as "certain knowledge," it played a central role in medieval Aristotelian discussions of method, where it referred to the demonstrative knowledge of essences. During the Renaissance and the seventeenth century, the growth of empirical investigation, especially in natural history, yielded less-than-certain knowledge about the world, making it increasingly difficult to assimilate natural philosophy into the Aristotelian apodictic model for science. Many natural philosophers developed an empiricist and probabilist theory of knowledge which further separated their endeavors from the Aristotelian model of *"scientia."*[13] The word was thus not new in the early modern period. Furthermore, early modern natural philosophers—who may or may not have been pursuing *scientia*—were not the same as modern scientists.[14] As late as 1690, John Locke could write "that natural philosophy is not capable of being made a science."[15] Although there were many sciences in the early modern period, there was not the general rubric Science that was to develop in a later period.[16]

But, to return to Schiebinger's analysis, is the feminine necessarily the object of desire, the opposite of the male natural philosopher? Many of the early modern philosophers who play an important role in this discussion would not have experienced a feminine *Natura* or *Philosophia* as an object of desire. Francis Bacon was homosexual, as probably was Isaac Newton. And Robert Boyle was, if not actively homosexual, then at least homosocial and, by his own testimony, celibate.[17]

Trading on feminine metaphors for nature, some scholars have found great significance in the rhetoric Bacon used to describe the experimental methods he advocated for the new natural philosophy.[18] As rendered by his nineteenth-century translator, Bacon wrote passages that appear to recommend harassment and torture as means of extracting nature's secrets. For example, in *Of the Dignity and Advancement of Learning,* Bacon wrote:

> For you have but to follow and as it were hound nature in her wanderings, and you will be able, when you like, to lead and drive her afterwards to the same place again. Neither am I of the opinion in this history of marvels, that superstitious narratives of sorceries, witchcrafts, charms, dreams, divinations, and the like, where there is an assurance and clear evidence of the fact, should be altogether excluded. For it is not yet known in what cases, and how far, effects attributed to superstition participate of natural causes; and therefore howsoever the use and practice of such arts is to be condemned, yet from the speculation and consideration of them (if they be diligently unraveled) a useful light may be gained, not only for the true judgement of the offences of persons charged with such practices, but likewise for the further disclosing of the secrets of nature. Neither ought a man to make scruple of entering and penetrating into these holes and corners, where the inquisition of truth is his sole object.[19]

Merchant interprets this passage as referring to the use of judicial torture by the Inquisition's persecution of witches and to "the supposed sexual crimes and practices of witches."[20]

Although the allusion to witchcraft is undeniable, reading the passage in context reveals that Bacon had very different objects in mind. Coming in the middle of his famous revision of the art/nature distinction as found in

Aristotle, Bacon noted that "it is nature that governs everything; but under nature are included these three; the *course* of nature, the *wanderings* of nature, and *art,* or nature with man to help; which three must therefore all be included in Natural History."[21] He then suggested that "from the wonders of nature is the most clear and open passage to the wonders of art."[22] Far from advocating imitation of the methods of the witch-hunters as a model for the experimental philosophy, Bacon was suggesting that the study of superstition and witchcraft—by following nature into these dark holes—would advance our knowledge of the natural world.[23]

The theme of torturing nature to extract her secrets rests on a careless translation of Bacon's Latin. In a close study of Bacon's linguistic practices, Peter Pesic argues that Bacon did not use the word "torture" *("tortura"* or *"quaestio")* in discussing the interrogation of nature.[24] Rather he used cognates of the verb *"vexare,"* "to vex": "The Latin root *vexare* suggests shaking, agitation, disturbance; the English uses of 'vexation' contemporaneous with Bacon pertain to conditions that are troubling, afflicting, or harassing. In many passages the mental trials inherent in vexation distinguish it from the sheer brutality of torture."[25] Bacon used the figure of Proteus, Neptune's immortal herdsman reputed in the *Odyssey* to know the past, present, and future, as an image of "the secrets of nature and the conditions of matter."[26] Proteus "has the power to take all manner of shapes, but if held till he resumes the true one, will answer questions."[27] It is necessary to wrestle with matter and even to bind it, just as Menelaos had had to wrestle with Proteus and bind him in handcuffs in order to extract information that the Greeks needed for the journey home from Troy. But wrestling implies a struggle between equals rather than the brutality and domination characteristic of torture.[28] As an approach to learning the secrets of nature, the use of vexations once again relates to the art/nature distinction and the relevance of human skill in uncovering truths about the world.

> Among the parts of history which I have mentioned, the history of Arts is of most use, because it exhibits things in motion, and leads more directly to practice. Moreover it takes off the mask and veil from natural objects, which are commonly concealed and obscured under the variety of shapes and external appearance. Finally, the vexations of art *[vexationes artis]* are certainly as the bonds and handcuffs of Proteus, which betray the ultimate struggles and efforts of matter. For bodies will not be destroyed or annihilated; rather than that they will turn themselves into various forms.[29]

Subjecting nature to art would force her to reveal secrets that would not emerge without human intervention.[30] Because *"ars"* is a feminine noun, these extended metaphors lack any notion of male domination.

In another passage that has aroused considerable attention from feminist scholars, Bacon wrote, "I am come in very truth leading to you Nature with all her children to bind her to your service and make her your slave."[31]

This sentence appears in *The Masculine Birth of Time,* an unfinished work that was not published in Bacon's lifetime. Since Bacon abandoned the project, it is probably a mistake to read too much significance into its details. Furthermore, as Sarah Hutton argues, the intellectual and rhetorical contexts of early modern metaphors were quite different from ours and cannot be interpreted in terms of twentieth- and twenty-first-century preoccupations.[32]

## The Organic/Mechanical Distinction

Another theme that permeates the feminist critique of science is the claim that the mechanical philosophy—which came to play a central role in seventeenth-century natural philosophy—represented a radical change in the concept of nature, one deleterious to the perception of the feminine. As a corollary to this development, Merchant claims that the organic view of nature embodied in alchemy was an alternative friendlier to women and feminine principles. Merchant writes:

> The mechanists transformed the body of the world and its female soul, source of activity in the organic cosmos, into a mechanism of inert matter in motion, translated the world spirit into a corpuscular ether, purged individual spirits from nature, and transformed sympathies and antipathies into efficient causes. The resultant corpse was a mechanical system of dead corpuscles, set into motion by the Creator, so that each obeyed the law of inertia and moved only by contact with another moving body.[33]

At least two themes implicit in this view must be revisited in light of recent scholarship: first, that the alchemical, animistic, "organic" view of nature was friendlier to women than the new mechanical philosophy; and second, that there existed a sharp demarcation between these two philosophies of nature.

Merchant's oft-quoted claim that the alchemical worldview embodied a more positive view of women and the feminine than that implicit in the mechanical philosophy is based on a selective set of readings from Paracelsus, chosen by Jolande Jacobi, one of Carl Jung's disciples, for the purpose of giving "modern man in his predicament that strength and faith which radiate from all creative natures and their works—to help the reader to gain new courage from the contemplation of the Paracelsian *compositio humana,* the nobility and dignity with which he endowed the concept of man."[34] Jacobi tells us that she purposely omitted many features of Paracelsus' writings that have "only secondary or historical significance" and have no bearing on her modern psychological and spiritual concerns.[35] This approach may be suitable for Jacobi's purposes, but it renders her book of questionable value as a source for the intellectual historian. Yet Jacobi's book is the source that both Merchant and feminist critic Evelyn Fox Keller used to support the claim that alchemy was friendlier to the feminine than the mechanical philosophy that destroyed it.[36]

William R. Newman, working from Paracelsus' original writings, tells a very different story about the Paracelsian conception of women, one that reeks of misogyny.[37] Newman describes Paracelsus' account of how to produce the basilisk and the homunculus, procedures which demonstrate the bounds beyond which human art cannot be pushed. Both procedures also reveal profound misogyny. Paracelsus wrote: "Now I return to my subject, to explain why and for what reason the basilisk has the poison in its glance and eyes. It must be known, then, that it has such a characteristic and origin from impure women. . . . For the basilisk grows and is born out of and from the greatest impurity of women, from the menses and the blood of the sperm."[38]

The production of the homunculus, basically a little man made in an alchemical flask, equally downgrades women. The homunculus is translucent, and it has wonderful powers. It can defeat its enemies and know all about the secrets of things. It has these characteristics because it is made by art from sperm, uncontaminated by contact with a woman or her menstrual blood.[39] Newman further demonstrates that the alchemical literature—in this case works by Zosimos of Panopolis and Arnald of Villanova (whom Merchant cites as examples of pro-feminine alchemical writers[40])—is larded with images of sadism and torture as methods for forcing nature to reveal her secrets.[41] Whether or not the mechanical and experimental philosophies can be exonerated from the charges of misogyny and torture, alchemy does not seem to offer a friendlier alternative.

The demarcation between the organic and mechanical philosophies of nature is not nearly as clear as Merchant and Keller supposed. Recent studies of early modern natural philosophy have revised the notion that the seventeenth century witnessed a sharp break from earlier animistic and organic views. There is now considerable evidence for lingering Aristotelianism, even within the thinking of the mechanical philosophers.[42] Corpuscular theories of matter developed from diverse roots, including Aristotelianism, alchemy, Paracelsianism, Epicureanism, and Stoicism.[43] From the outset, mechanical philosophers were unable to solve the problems they addressed in the strictly mechanical terms of dead matter and a mechanics based on inertial motion. In addressing this problem, they drew on diverse organic philosophies of nature, including alchemy, Renaissance naturalism, and even Aristotelianism.[44] Moreover, many of the most prominent mechanical philosophers had an abiding interest in and knowledge of alchemy, which they managed to incorporate into their broader philosophy of nature.[45]

## Was the Gender of Nature a Significant Issue for Early Modern Natural Philosophers?

What, if anything, did early modern natural philosophers think about the gender of nature? A salient example is Robert Boyle (1627–1691), who devoted an entire treatise to the concept of nature. Boyle composed *A Free Enquiry into the Vulgarly Receiv'd Notion of Nature* (1686) in the 1660s at a time when he was articulating the general principles of his corpuscularian philosophy.[46] Boyle's work, as a whole, can be best understood in the context of the seventeenth-century quest for a philosophy of nature to replace the Aristotelianism that had lost credibility in the wake of the Copernican Revolution, the skeptical crisis, the Protestant Reformation, and the humanist revival of alternative ancient philosophies of nature. By the middle of the seventeenth century, there was an active competition among several of these philosophies of nature.[47] The mechanical philosophers, influenced by René Descartes (1596–1650) and Pierre Gassendi (1592–1655), advocated a mechanical analogy to explicate the phenomena of nature.[48] Followers of Paracelsus espoused a chemical philosophy, which endeavored to account for natural phenomena in terms of a more holistic, chemical metaphor.[49] And, of course, there remained firmly entrenched Aristotelians, especially in the universities where the curriculum was based on the philosophy of Aristotle and his Scholastic commentators.[50] The points of contention among these groups were as much theological as they were scientific.

From his earliest writings on natural philosophy (1649–1653), Boyle favored the mechanical philosophy, although he also borrowed freely from the other traditions. Drawing on both Gassendi's Christianized version of Epicurean atomism and Descartes' mechanical philosophy and departing from each in certain crucial ways, Boyle elaborated his own "corpuscular hypothesis" to explain the phenomena of the physical—and most especially chemical—world.[51] The most fundamental assertion of Boyle's corpuscularianism—and every other version of the mechanical philosophy—is that all physical (that is, not spiritual) phenomena can be explained in terms of matter and motion. All the different kinds of matter encountered in the world can be reduced to "one catholick or universal matter common to all bodies, by which I mean substance, extended, divisible, impenetrable."[52]

Since matter is inactive, it cannot be the source of diversity and change in the world. Another principle, motion, is needed.[53] God is the source of all the motion in the world. He set matter into motion when he created it. Material objects are composed of particles which lie below the threshold of sense and which, divisible in principle, are hardly ever divided. Boyle called these particles *"minima naturalia."* The *minima naturalia* form clusters that combine to form macroscopic bodies. The shapes, sizes, and motions of the clusters and their constituent *minima* cause the qualities and changes observed in macroscopic objects.[54] Different kinds of substances are formed by different arrangements of the particles composing the clusters that unite to form bodies. And changes in bodies occur when the composition or arrangement of the clusters changes. Given the ontological framework of the mechanical world, Boyle then sought a theoretical account of the chemical substances and processes that he understood from his laboratory experience.

While Boyle's scientific writings were largely devoted to providing support for his corpuscularianism and for applying it to chemical phenomena, like almost all the other seventeenth-century mechanical philosophers,

Boyle found himself having to defend his views against charges that natural philosophy in general would become a serious distraction from religion and that the mechanical philosophy in particular would lead to atheism and materialism. In articulating his corpuscularian philosophy and defending it against charges of materialism and atheism, Boyle developed a complex position that provided philosophical and theological foundations for his philosophy of nature. These foundations consisted of a voluntarist understanding of God's relationship to the creation, a nominalist ontology, and an empiricist approach to scientific knowledge.

Central to Boyle's theology, metaphysics, and natural philosophy was his solidly voluntarist conception of God's relationship to the creation. According to Boyle, everything in the world is utterly contingent on divine will. The laws of nature are what they are because God created them so; it is not the case that he created them because they are true. He can alter them at will.[55] Miracles are evidence of his acting freely, since nothing, not even the laws of nature, can obstruct God from exercising his will freely.[56] Boyle repeatedly alluded to God's freedom in creating the world and to his power over the laws of nature: "the laws of nature, as they were at first arbitrarily instituted by God, so, in reference to him, they are but arbitrary still."[57]

These theological concerns dominated his discussion in the *Notion of Nature*. In that book, he set out to examine what the word "nature" meant in contemporary natural philosophy. "The word 'nature' is everywhere to be met with in the writings of physiologers [natural philosophers]. But though they frequently employ the word, they seem not to have much considered what notion ought to be framed of the thing, which they suppose and admire, and upon occasion celebrate, but do not call in question to discuss."[58]

What is the vulgarly received notion of nature? Boyle noted that the term "nature" was frequently employed, although the meaning of the term remained confused. Early in the *Notion of Nature* he enumerated various usages of the term in common parlance.

> The vulgar notion of nature may be conveniently enough expressed by some such description as this.
>
> Nature is a most wise being that does nothing in vain, does not miss of her ends, does always that which (of the things she can do) is best to be done, and this she does by the most direct or compendious ways, neither employing any things superfluous, nor being wanting in things necessary; she teaches and inclines every one of her works to preserve itself. And, as in the microcosm (man) it is she that is the curer of diseases, so in the macrocosm (the world) for the conservation of the universe she abhors a vacuum, making particular bodies act contrary to their own inclinations and interests to prevent it for the public good.[59]

In using the feminine pronoun here to refer to nature, Boyle was alluding to the practice of others who anthropomorphized nature, a practice he disdained because anthropomorphizing the concept resulted in its reification.[60]

Boyle objected to the reification of nature, just as he objected to the reification of Fortune, which "the Gentiles" had made into a goddess and which "eminent writers in verse and in prose, ethnic and Christian, ancient and modern, and all sorts of men in their common discourse do seriously talk of it as if it were some kind of Antichrist that usurped a great share in the government of the world."[61] The main reasons for his objection were theological. There is no mention of "nature" in the Bible, and reification of the concept contradicted Boyle's conception of the deity.[62]

Consistent in his voluntarist conception of God's relationship to the creation, Boyle was an anti-essentialist, and anti-essentialism directly informed both his philosophy of nature and his scientific method. Anti-essentialism in this context meant the denial of reality to essences or natures. All that exists in Boyle's world are individuals. There are no Platonic forms or Aristotelian natures. Accordingly, he regarded his corpuscularianism as entirely consistent with his voluntarist theology because it did not allow for the real existence of essences or universals. Such entities would impede God's free exercise of his will. Boyle argued that the philosophies of nature associated with Aristotelians, Platonists, Hermeticists, and Paracelsians did not allow enough scope for divine will because they interposed other, unnecessary orders of being between God and his creation[63]: "For ought [*sic*] I can clearly discern, whatsoever is perform'd in the merely material world, is really done by particular bodies, acting according to the laws of motion, rest, &c. that are settled and maintained by God among things corporeal."[64] Among the agents intermediate between God and the created world that Boyle explicitly rejected were the Chance of the Epicureans and the Fate of the Stoics, as well as the "vulgarly received notion of nature." He rejected them because they would interfere with the free exercise of divine will, by presupposing the existence of some kind of agent mediating between God and the creation. Boyle considered chance to be a poor substitute for providence, the manifestation of divine wisdom and power.[65] Similarly, fate was an illegitimately reified notion, expressing no more than the causal nexus created world.[66]

Reification is bad enough; thinking of nature as some kind of goddess amounts to idolatry, a practice that Boyle denounced for many pages in the *Notion of Nature*.[67] Invoking the principle of parsimony, Boyle stated, "I see not why we should take a principle of which we have no need."[68] Indeed, he argued, his natural philosophy—corpuscularianism combined with God as divine artificer—suffices to explain everything that the "naturists" ascribe to their otiose concept of nature.

> For supposing the common matter of all bodies to have been at first divided into innumerable minute parts by the wise author of nature, and these parts to have been so disposed of as to form the world, constituted as it now is; and especially, supposing that the universal laws of motion among the parts of the matter have been established, and several conventions of particles contrived into the seminal principles of various things; all which may be effected by the mere local motion

of matter (not left to itself, but skillfully guided at the beginning of the world)—if (I say) we suppose these things, together with God's ordinary and general concourse, which we very reasonably may, I see not why the same phenomena that we now observe in the world should not be produced, without taking in any such powerful and intelligent being, distinct from God, as nature is represented to be. And till I see some instance produced to the contrary, I am like to continue of this mind and to think that the phenomena we observe will genuinely follow from the mere fabric and constitution of the world.[69]

Boyle's analysis of the concept of nature neither drew on nor perpetuated the misogynist tendencies ascribed to Bacon's metaphors and the mechanical philosophy. Far from it. Boyle rejected the anthropomorphizing of nature and redefined the concept of nature in terms of his physico-theological project. The world consists of matter, created, designed, and moved by God, who is the only source of motion and order in the universe. Nature is neither a nurturing mother nor a woman to be hounded and tortured to reveal her secrets. Boyle—who was one of the most prominent natural philosophers of the second half of the seventeenth century—generally used metaphors that are neutral on the subject of the gender of nature.[70]

## Why Were Women Excluded from Early Modern Natural Philosophy?

That women were excluded from natural philosophy during the early modern period is well documented. Attempts to blame this exclusion on the metaphors used to describe nature and the methods of science have run aground on the shoals of careful textual analysis. Metaphors may reflect social realities, but I seriously doubt that they cause social relations. Instead of looking at metaphors about the gender of nature to explain the undisputed facts of the exclusion of women from natural philosophy, I would be inclined to look at the history of the institutions and societies in which natural philosophy was practiced in order to learn about the sources of their particular social practices.

The path is fairly well marked. Natural philosophy grew from the combination of Greek philosophy and Christian theology as pursued within medieval universities that were founded as clerical institutions.[71] The medieval universities—like the Pythagorean community, Plato's Academy, and Aristotle's Lyceum—by and large excluded women, and that tradition of exclusion continued into the early modern period and beyond, with only a few exceptions.[72] Natural philosophy also developed in the courts of Europe where women often played active but still marginal roles in intellectual life.[73] Early modern universities and scientific societies were frequently modeled on these earlier institutions and continued to be homosocial and in some cases homosexual. These exclusionary practices were rarely questioned at the time.

Rather than focusing on the gender of nature, historians should examine the nature of gender and gender relations in order to understand why women were excluded from early modern natural philosophy. Recent scholarship—to which the present volume bears witness—is taking giant steps in this direction. Historians of science are considering gender as well as other areas that traditionally have been ignored. They are discovering hitherto unknown women who actively pursued knowledge of and careers in the sciences, and they are studying the social conditions that either prevented or enabled women to engage in these pursuits. These recent studies are providing a more historically contextualized and nuanced understanding of how gender functioned in the social relations of early modern sciences.[74]

## Notes

Acknowledgments: I am grateful to Sarah Hutton and Margaret G. Cook for helpful suggestions on an earlier version of this paper.

1. See Carolyn Merchant, *The Death of Nature: Women, Ecology, and the Scientific Revolution* (San Francisco: Harper and Row, 1980); Evelyn Fox Keller, *Reflections on Gender and Science* (New Haven: Yale University Press, 1985); and Susan R. Bordo, *The Flight to Objectivity: Essays on Cartesianism and Culture* (Albany: SUNY Press, 1987).

2. See Margaret J. Osler, "The Canonical Imperative: Rethinking the Scientific Revolution," in *Rethinking the Scientific Revolution,* ed. Margaret J. Osler (Cambridge: Cambridge University Press, 2000), 3–22.

3. See, for example, Genevieve Lloyd, *The Man of Reason: "Male" and "Female" in Western Philosophy* (Minneapolis: University of Minnesota Press, 1984); David Noble, *A World without Women: The Christian Clerical Culture of Western Science* (New York: Knopf, 1992), 222–24; Margaret Wertheim, *Pythagoras' Trousers: God, Physics, and the Gender Wars* (New York: Norton, 1995), chap. 4; and John Rogers, *The Matter of Revolution: Science, Poetry, and Politics in the Age of Milton* (Ithaca: Cornell University Press, 1996).

4. Merchant, *The Death of Nature,* 168–72.

5. Ibid., chap. 8.

6. Ibid., chap. 4.

7. Ibid., introduction and chap. 1. Merchant states: "Both the nurturing and domination metaphors had existed in philosophy, religion, and literature. The idea of dominion over the earth existed in Greek philosophy and Christian religion; that of the nurturing earth, in Greek and other pagan philosophies. But, as the economy became modernized and the Scientific Revolution proceeded, the dominion metaphor spread beyond the religious sphere and assumed ascendancy in the social and political spheres as well." Ibid., 3.

8. Ibid., xix.

9. For technical discussions of the concept of gender in linguistics, see Dennis Baron, *Gender and Grammar* (New Haven: Yale University Press, 1986); and Greville Corbett, *Gender* (Cambridge: Cambridge University Press, 1991). I am grateful to Betsy Ritter for providing me with these references.

10. Sarah Hutton makes a similar point in "The Riddle of the Sphinx: Francis Bacon and the Emblems of Science," in *Women, Science and Medicine, 1500–1700: Mothers and Sisters of the Royal Society,* ed. Lynette Hunter and Sarah Hutton (Phoenix Mill Thrupp: Sutton, 1999), 7–28.

11. Londa L. Schiebinger, *The Mind Has No Sex? Women in the Origins of Modern Science* (Cambridge, MA: Harvard University Press, 1989), 134.

12. Ibid.

13. Barbara J. Shapiro, *Probability and Certainty in Seventeenth-Century England: A Study of the Relationships Between Natural Science, Religion, History, Law, and Literature* (Princeton, NJ: Princeton University Press, 1983); Ernan McMullin, "Conceptions of Science in the Scientific Revolution," in *Reappraisals of the Scientific Revolution,* ed. David C. Lindberg and Robert S. Westman (Cambridge: Cambridge University Press,

1990), 27–92; Margaret J. Osler, "John Locke and the Changing Ideal of Scientific Knowledge," *Journal of the History of Ideas* 31 (1970), 3–16 (reprinted in *Philosophy, Religion and Science in the 17th and 18th Centuries,* ed. John W. Yolton [Rochester: University of Rochester Press, 1990], 325–38); Margaret J. Osler, *Divine Will and the Mechanical Philosophy: Gassendi and Descartes on Contingency and Necessity in the Created World* (Cambridge: Cambridge University Press, 1994), chap. 4; Jan W. Wojcik, *Robert Boyle and the Limits of Reason* (Cambridge: Cambridge University Press, 1996).

14. Sydney Ross, "*Scientist:* The Story of a Word," *Annals of Science* 18 (1962), 65–85.

15. John Locke, *An Essay Concerning Human Understanding,* ed. Peter H. Nidditch (Oxford: Clarendon Press, 1975), bk. 4, chap. 12, §10, 645.

16. A. Cunningham, "Getting the Game Right: Some Plain Words on the Identity and Invention of Science," *Studies in History and Philosophy of Science* 19 (1988), 365–89.

17. On the question of Bacon's sexuality, see Perez Zagorin, *Francis Bacon* (Princeton: Princeton University Press, 1998), 12–14. Newton claimed to have died a virgin, but his attraction to Fatio de Duillier is well documented. See Richard S. Westfall, *Never at Rest: A Biography of Isaac Newton* (Cambridge: Cambridge University Press, 1980), 59, 495–97, 528–33. Boyle remained celibate throughout his life, a fact known in his lifetime and from his own testimony. See Michael Hunter, "Alchemy, Magic and Moralism in the Thought of Robert Boyle," *British Journal for the History of Science* 23 (1990), 319; and Michael Hunter, *Robert Boyle by Himself and His Friends* (London: William Pickering, 1994), lxxvii–lxxviii.

18. Frequently cited examples are Merchant, *The Death of Nature,* 168–72; and Keller, *Reflections on Gender and Science,* chap. 2.

19. Francis Bacon, *Of the Dignity and Advancement of Learning,* in *The Works of Francis Bacon,* ed. James Spedding, Robert Leslie Ellis, and Douglas Denon Heath, 14 vols. (London: Longman, 1858), 4:296.

20. Merchant, *The Death of Nature,* 168–69.

21. Bacon, *Dignity and Advancement of Learning,* 295.

22. Ibid., 296. Lorraine Daston and Katharine Park highlight the historical significance of Bacon's advocacy of studying wonders and marvels as an important part of natural history. See their *Wonders and the Order of Nature: 1150–1750* (New York: Zone Books, 1998), 220–31.

23. This point is more fully developed in Alan Soble, "In Defense of Bacon," *Philosophy of the Social Sciences* 25 (1995), 192–215. Soble's paper was augmented and republished in *A House Built on Sand: Exposing Postmodernist Myths about Science,* ed. Noretta Koertge (New York: Oxford University Press, 1998), 195–215.

24. Peter Pesic, "Wrestling with Proteus: Francis Bacon and the 'Torture' of Nature," *Isis* 90 (1999), 81–94.

25. Ibid., 88–89.

26. Ibid., 84.

27. N. G. L. Hammond and H. H. Scullard, eds., *The Oxford Classical Dictionary,* 2nd ed. (Oxford: Clarendon, 1970), 901.

28. On the use of torture in the English courts of law, see Clifford Hall, "Some Perspectives on the Use of Torture in Bacon's Time and the Question of his 'Virtue'," *Anglo-American Law Review* 18 (1989), 289–321.

29. Bacon, *Preparative towards a Natural and Experimental History,* in *Works,* 4:257 (*Paracsceve ad Historiam Naturalem et Experimentalem* [1620], vol. 1, 398–99).

30. Keller argues for Bacon's misogyny in terms similar to Merchant's but focuses on his unpublished treatise, *The Masculine Birth of Time.* For a critique of Keller and an alternative analysis of Bacon's metaphor, see Hutton, "The Riddle of the Sphinx."

31. Francis Bacon, *The Masculine Birth of Time, or Three Books on the Interpretation of Nature,* in Benjamin Farrington, *The Philosophy of Francis Bacon* (Chicago: University of Chicago Press, 1966), 62.

32. Hutton, "The Riddle of The Sphinx," 13–21.

33. Merchant, *The Death of Nature,* 195.

34. Jolande Jacobi, preface to *Paracelsus, Selected Writings,* ed. Jolande Jacobi (New

York: Bollingen, 1951), 34.

35. Ibid., 27.

36. Merchant, *The Death of Nature,* 298n19ff.; Keller, *Reflections on Gender and Science,* 43, 49, 52, 53. For a critique of the Jungian interpretation of alchemy, see Lawrence M. Principe and William R. Newman, "Some Problems with the Historiography of Alchemy," in *Secrets of Nature: Astrology and Alchemy in Early Modern Europe,* ed. William R. Newman and Anthony Grafton (Cambridge, MA: MIT Press, 2001), 385–431; and Richard Noll, *The Jung Cult: Origins of a Charismatic Movement,* new ed. (New York: Free Press, 1997).

37. William R. Newman, "Alchemy, Domination, and Gender," in Koertge, *A House Built on Sand,* 216–26.

38. Paracelsus [pseud.], *De rerum natura,* in Karl Sudhoff (1928), 11:315, as quoted and translated by Newman in "Alchemy, Domination, and Gender," 219.

39. Ibid., 219–20.

40. Merchant, *The Death of Nature,* 18, 22.

41. On the misogyny of the alchemists, see also Mary Tiles, "Mathesis and the Masculine Birth of Time," *International Studies in the Philosophy of Science* 1 (1986), 16–35.

42. Margaret J. Osler, "New Wine in Old Bottles: Gassendi and the Aristotelian Origin of Early Modern Physics," *Midwest Studies in Philosophy,* special issue on *Renaissance and Early Modern Philosophy* 26 (2002), 167–84.

43. See Christoph Lüthy, John Murdoch, and William R. Newman, eds., *Late Medieval and Early Modern Corpuscular Matter Theory* (Leiden: Brill, 2001).

44. See John Henry, "Occult Qualities and the Experimental Philosophy: Active Principles in Pre-Newtonian Matter Theory," *History of Science* 24 (1986), 335–81; Margaret J. Osler, "How Mechanical Was the Mechanical Philosophy? Non-Epicurean Themes in Gassendi's Atomism," in *Late Medieval and Early Modern Corpuscular Matter Theory,* 423–39; and Antonio Clericuzio, *Elements, Principles and Corpuscles: A Study of Atomism and Chemistry in the Seventeenth Century* (Dordrecht: Kluwer, 2000).

45. Boyle and Newton are the most notable examples, although there are many others. See Lawrence M. Principe, *The Aspiring Adept: Robert Boyle and His Alchemical Quest* (Princeton: Princeton University Press, 1998); B. J. T. Dobbs, *The Foundations of Newton's Alchemy or, 'The Hunting of the Greene Lyon'* (Cambridge: Cambridge University Press, 1975); and B. J. T. Dobbs, *The Janus Faces of Genius: The Role of Alchemy in Newton's Thought* (Cambridge: Cambridge University Press, 1991).

46. Michael Hunter and Edward B. Davis, "The Making of Robert Boyle's *Free Enquiry into the Vulgarly Receiv'd Notion of Nature* (1686)," *Early Science and Medicine* 1 (1996), 204–71.

47. See Richard H. Popkin, *The History of Scepticism from Erasmus to Spinoza* (Berkeley: University of California Press, 1979).

48. See Osler, *Divine Will and the Mechanical Philosophy.*

49. See Allen G. Debus, *Science and Education in the Seventeenth Century: The Webster-Ward Debate* (New York: Neale Watson, 1970).

50. See Dennis Des Chene, *Natural Philosophy in Late Aristotelian and Cartesian Thought* (Ithaca: Cornell University Press, 1996); and Roger Ariew, *Descartes and the Last Scholastics* (Ithaca: Cornell University Press, 1999).

51. See Marie Boas Hall, *Robert Boyle and Seventeenth-Century Chemistry* (Cambridge: Cambridge University Press, 1958); Michael Hunter, ed., *Robert Boyle Reconsidered* (Cambridge: Cambridge University Press, 1994); and Principe, *The Aspiring Adept.*

52. Robert Boyle, *The Origine of Formes and Qualities (according to the Corpuscular Philosophy) Illustrated by Considerations and Experiments,* in *The Works of Robert Boyle,* ed. Michael Hunter and Edward B. Davis, 14 vols. (London: Pickering and Chatto, 1999–2000), 5:305.

53. Ibid.

54. Ibid., 326.

55. Robert Boyle, *Some Considerations about the Reconcileableness of Reason and Religion,* in *Works,* 8:251–52.

56. "It is a rule in natural philosophy that *causae necessariae semper agent quantum possunt;* but it will not follow from thence, that the fire must necessarily burn *Daniels*

three companions . . . , when the author of nature was pleased to withdraw his concourse to the operation of the flames, or supernaturally to defend against them the bodies, that were exposed to them." Ibid., 252.

57. Robert Boyle, *An Appendix to the First Part of The Christian Virtuoso,* in *Works,* 11:423.

58. Robert Boyle, *A Free Enquiry Into the Vulgarly Receiv'd Notion of Nature; Made in an Essay Address'd to a Friend,* in *Works,* 10:439.

59. Ibid., 463.

60. Boyle was not entirely consistent and occasionally used anthropomorphic metaphors to describe nature. For example, in *Certain Physiological Essays,* 2nd ed. (1669), Boyle wrote, ". . . there are two distinct Ends that Men may propound to themselves in studying Natural Philosophy. For some Men care only to Know Nature, others desire to Command Her: or to express it otherwise, some there are who desire but to Please themselves by the Discovery of the Causes of the known Phænomena, and others would be able to produce new ones, and bring Nature to be serviceable to their particular Ends, whether of Health, or Riches, or sensual Delight" (in *Works,* 2:23). Given the position he fully articulated in the *Notion of Nature,* this usage seems to be a rhetorical cliché rather than a reflection of his genuine point of view.

61. Boyle, *Notion of Nature,* 451.

62. Ibid., 446–47

63. Ibid., 522.

64. Ibid., 499–500.

65. Ibid., 530–32.

66. Ibid., 109.

67. Ibid., 473–83. "And however, the great and pernicious errors they were led into by the belief that the universe itself and many of its nobler parts besides men were endowed not only with life, but understanding and providence, may suffice to make us Christians very jealous of admitting such a being as that which men venerate under the name of 'nature', since they ascribe to it as many wonderful powers and prerogatives as the idolaters did to their adored mundane soul." Ibid., 478.

68. Ibid, 478.

69. Ibid.

70. In a recent study, Elizabeth Potter discusses Boyle's attitudes towards women and the social context of his experiments with the air pump in order to establish that Boyle's Law was not founded on strictly *scientific* (in the modern sense) grounds. Although she gives some attention to his theology, she does not provide any analysis of his concept of nature. See Elizabeth Potter, *Gender and Boyle's Law of Gases* (Bloomington: Indiana University Press, 2001).

71. See Edward Grant, *The Foundations of Modern Science in the Middle Ages: Their Religious, Institutional, and Intellectual Contexts* (Cambridge: Cambridge University Press, 1996); and Roger French and Andrew Cunningham, *Before Science: The Invention of the Friars' Natural Philosophy* (Brookfield, VT: Scolar Press, 1996).

72. There are a few notable exceptions, the most famous of which is Margaret Cavendish. See Schiebinger, *The Mind Has No Sex?* 47–59. See also Tara E. Nummedal, "Alchemical Reproduction and the Career of Anna Maria Zieglerin," *Ambix* 48 (2001), 56–68; and Lucia Tosi, "Marie Meurdrac: Paracelsian Chemist and Feminist," *Ambix* 48 (2001), 69–82.

73. Schiebinger, *The Mind Has No Sex?* 44–47.

74. Historians are undertaking new studies that explore these directions. Schiebinger's *The Mind Has No Sex?* raises questions that point to a number of fruitful areas for further research, many of which have been taken up since the publication of her work. See Paula Findlen, "Science as a Career in Enlightenment Italy: The Strategies of Laura Bassi," *Isis* 84 (1993), 441–69; Paula Findlen, "Translating the New Science: Women and the Circulation of Knowledge in Enlightenment Italy," *Configurations: A Journal of Literature, Science, and Technology* 3 (1995), 167–203; Mary Terrall, "Émilie du Châtelet and the Gendering of Science," *History of Science* 33 (1993), 283–310; and Mary Terrall, "Salon, Academy, and the Boudoir: Generation and Desire in Maupertuis's Science of Life," *Isis* 87 (1996), 217–29.

J. B. SHANK

# Neither Natural Philosophy, Nor Science, Nor Literature

## *Gender, Writing, and the Pursuit of Nature in Fontenelle's* Entretiens sur la pluralité des mondes habités

Accounts of the making of modern science still have not fully escaped from the creation stories of the so-called Scientific Revolution. Nowhere is this more true than in writing about European natural knowledge during the years 1670–1720. This half century witnessed the so-called Newtonian Synthesis,[1] the event which A. R. Hall once called, in a canonical formulation, "the climax [of the Scientific Revolution] so far as the physical sciences are concerned."[2] After Newton, Hall continued, "the astronomical and cosmological issues that had so troubled the world since Copernicus's time were regarded as settled for good; it only remained for mathematicians to arrange the details of the Newtonian universe in somewhat more exact order."[3] Triumphalist stories such as these once defined research in the history of early modern science, yet their credibility today has all but evaporated. In the wake of their disintegration, we are now confronted with a new problem: replacing the founding mythology of the history of science with new genealogical accounts that give greater emphasis to historical complexity, local context, and developmental contingency.

De-centering Newton from the history of this period has been one salutary move since it has effected an historiographical escape from the all-consuming funnel narrative of the "Newtonian Revolution."[4] Yet since Newton's importance in the making of modern science cannot be ignored, the ongoing project of historically deconstructing his work and legacy has been crucial as well.[5] These efforts must be joined, however, with still more attempts to expand and historically complicate the contextual lens through which we understand Newton's place within the wider culture of European knowledge production. In particular, Newton's work must be considered not only in terms of the successful modern sciences that it spawned but also in relation to the other trajectories inherent in the Newtonian project—the union of physics and theology, for example—that history has chosen to cast into the dustbin. This research must be aggressive in escaping the teleological biases of the classic historiography by demonstrating how these alternatives constituted something other than wrong turns in the natural development of the true science of physics. The last point is particularly important from a cultural history point of view, for if it has now become a truism that the transformation of early modern natural philosophy into modern science brought with it a corresponding transformation in social and political life, then historically reconstituting the lost alternatives to Newtonian physics can also open up perspectives on the lost social and political possibilities of the period as well.

It is this final set of agendas that I want to explore in this essay. Focusing on the particular case of French natural knowledge in the decades before and after the publication of Newton's *Principia,* I want to reconstitute the historical alternative which it offered to the "Newtonian way (or ways)" in science. In so doing, I particularly want to emphasize the distinctive gender character of this lost approach to nature and the cultural consequences that followed from its emergence and ultimate disappearance. Like historians of science more generally, feminist scholars concerned with the historical development of gender and science have remained trapped for too long in the canonical meta-narratives of the Scientific Revolution. Consequently, while important work has been done on the gender structures inherent in modern science, feminist scholars have too often allowed teleological assumptions about the "revolutionary birth" of modern science to determine their accounts of female exclusion and male domination in this sphere. As I have said, accounts of the historical emergence of modern science need to emphasize the contingency and complex historicity of these developments, and narratives about the gendering of science, I would argue, profit from emphases as well.

To study the case of France, I will focus my analysis on the work of Bernard le Bovier de Fontenelle, a writer, philosopher, man of letters, and mathematician who served as the Perpetual Secretary of the French Royal Academy of Sciences from 1697 to 1740. Fontenelle has figured centrally in contemporary accounts of the gendering of science, including those offered by Londa Schiebinger, Erica Harth, Mary Terrall, and others. Yet the full historical significance of his philosophical work remains to be constituted since his texts have not yet been fully embedded within the seventeenth-century categories that gave them meaning.[6] Nina Rattner Gelbart charts the right course when she warns us "not to fall into the trap of conceiving of Fontenelle as a bridge between two cultures, one scientific, the other humanistic." Despite his "matchless talent as a communicator and an interpreter," she asserts, we should not picture "experts producing knowledge and Fontenelle diluting it for popular consumption." Rather, the author of the *Entretiens sur la pluralité des mondes habités,* Fontenelle's most influential work, gives us invaluable insight into the world that preceded the complete birth and institutionalization of modern science, a time when "science was still in its adolescence, still searching for its purpose and its self-image, still seeking a public to understand it, make it welcome, foster and even guide it."[7]

Emma Spary's work on eighteenth-century French natural history and Jessica Riskin's work on what she calls the "sentimental empiricism" of the eighteenth century are two very recent steps in Gelbart's direction.[8] They also emphasize the absence of any strong separation between science and literature, or between reason and sentiment, in the eighteenth century. Consequently, they urge historians to "recover the links that existed between taste and reason, connoisseurship and utility, sensibility and science" in this period.[9] In this essay I would like to follow the lead of Gelbart, Spary, and Riskin by reconstituting the French natural knowledge that Fontenelle personified. As we will see, it was a philosophy that lay somewhere between traditional early modern natural philosophy and eighteenth-century Newtonian science, and it was a philosophy which likewise offered a gender alternative to the excessively patriarchal and masculine philosophical cultures that both preceded and succeeded it.

## Fontenelle's *Les Mondes* and the Institutions of French Natural Knowledge around 1680

Few works were as influential in shaping intellectual life in fin de siècle France as Fontenelle's *Entretiens sur la pluralité des mondes habités,* which first appeared in 1686.[10] *Les mondes,* as it affectionately came to be called, was an instant classic, passing through thirty-three editions and abundant translations by the middle of the eighteenth century. It also exerted a singular influence on French philosophical discourse.[11] Indeed, even a century after its publication, the work was still directly shaping philosophical writing both in France and in the wider international Republic of Letters.[12] Documenting the astonishing success of *Les mondes* is thus an easy undertaking; trying to account for it intellectually and culturally, however, is much more difficult. Part of the problem stems from the fundamental difficulty of simply categorizing the book. Gelbart's warning has largely gone unheeded. Virtually all scholars today implicitly or explicitly adopt a "Two Cultures" framework when describing Fontenelle's masterpiece.[13] They routinely characterize it as a "popularization of Cartesian science" designed to bridge the world of technical natural philosophy with the allegedly separate world of popular literary culture. Adopting the gender relations that are said to correspond to these two distinct spheres, the text's narrative structure—a dialogue between a learned man and an interested Marquise—is further read as the natural offspring of the text's position between male science on the one hand and female literary culture on the other, a dichotomy that also assumes a natural divide between masculine reason and feminine sentiment.

Yet to naturalize this understanding of the book is to impose an anachronistic set of divisions onto a world that was driven by a very different set of priorities. A better approach involves historically reconstructing the discursive and genre assumptions that made possible Fontenelle's writing in *Les monde.* The easiest place to begin is through a literal, if historically careful, description of the text itself.

When it appeared in 1686, Fontenelle's text contained a "Preface" by the author, an introductory "Letter à M. L....," which set up the subsequent text, and five consecutive conversations, each called a *"soir,"* or night, which were numbered in sequential order. After the author's preface, which described the goals of the book, the opening letter introduced the dialogues by claiming that they were the verbatim report, requested by M. L..., of what transpired during the narrator's recent visit to Madame la Marquise de G...'s country house. So introduced, the five *soirs* which followed offered a set of conversations between the narrator, an anonymous learned man, and the Marquise, an intelligent, aristocratic lady, which were said to have been held over five consecutive evenings in the garden near the Marquise's home. The dialogues themselves quickly let go of the framing device that introduced them, and once under way, they naturally positioned the reader as an eavesdropper upon a set of private conversations which were sometimes serious, frequently witty, occasionally flirtatious, and always lively.[14]

As the title suggests, the conversations in *Les mondes* center on the possibility that other worlds exist in the universe. They further explore whether these worlds, should they exist, could contain creatures and civilizations comparable to our own. For this reason, *Les mondes* often figures in historical accounts of the emergence of the modern extraterrestrial debate.[15] In the course of the discussions, however, a great deal of late-seventeenth-century cosmology is also explained, and since the cosmology that Fontenelle offers is the vortical-mechanical account of the universe first developed by René Descartes, it has become more common to describe *Les mondes* as a popularization of Cartesian science.

Certainly this characterization is not without merit. Cartesian cosmology was the leading scientific framework of the time—hence Newton's need to refute it directly in his *Principia* of 1687—and Fontenelle's 1686 text does offer an accurate, accessible, and highly readable account of this cosmology appropriate even for beginners. Yet to move from this fact to a characterization of *Les mondes* as, in essence, a "popularization" is to assume without scrutiny a set of assumptions about the nature of scientific inquiry, texts, institutions, and actors that are deeply anachronistic for the late seventeenth century.

What kind of text was *Les mondes* when considered in relation to other works of the late seventeenth century? Fontenelle addressed this question in his "Preface," noting that his was not an ordinary work easily categorized according to the intellectual expectations of the period. As he described the project: "I wanted to treat philosophy in a manner that was not at all philosophical; I tried to bring it to the point where it was neither too dry for the *gens du monde,* nor too frivolous *[badine]* for savants."[16] The author invoked Cicero's translation of Greek philosophy into the Latin vernacular of the Roman Republic as his inspiration, and in this way he positioned *Les mondes* as a text straddling two worlds, one learned and the other not. He also suggested that his book was an attempt to unite these two worlds, worrying at the outset whether in trying to construct a "middle ground" where learned and popular readers might meet he had in fact produced a text that would push them farther apart. Be that as it may, Fontenelle's language suggests a synthesis of the learned and the popular rather than a translation (read: popularization) of learned culture into the language of the wider public. To understand what Fontenelle might have had in mind, then, let us consider the expectations that each of the groups Fontenelle hoped to wed—savants and *gens du monde*—might have brought to their respective reading of the text.

On one side of the aisle were "savants," a very broad and contested term in late-seventeenth-century France. Who were the savants to which Fontenelle referred and how might they have been expected to read *Les mondes?* One group which obviously fell into this category were the *collège* and university professors whose authority, power, and prestige remained enormous in seventeenth-century France. The University of Paris continued to be the center of official scholastic culture, possessing a monopoly over education at the highest level (i.e., the granting of doctorates of philosophy, which were required of clerical authorities) and enjoying close ties with the French crown through its supervision of censorship and religious orthodoxy.[17] Yet its dominance in the intellectual sphere was not absolute. Jesuit *collèges* were entering their second century of existence in France when Fontenelle's text appeared, and by the late seventeenth century they had become a major force in French higher education, especially catering to elites who wanted a rigorous, modern curriculum for their sons destined for professional careers.[18] In fact, Fontenelle's parents, who were of middling noble rank, sent their son to a Jesuit *collège* in Normandy hoping that he would pursue a career in the law.[19] Other teaching orders also emerged in France alongside the Jesuits—the Oratorians were most important from a scientific point of view—and the existence of so many new educational institutions in the seventeenth century points to the perception, shared by many, that French higher education was undergoing transformation.[20]

Given the complexity of the French scholastic establishment, the word "savant" could be applied with equal accuracy to a deeply conservative Sorbonne theologian or to a *mondain* Jesuit philosopher. Nevertheless, those associated with the schools, whatever their stripe, did constitute a distinctive subculture among savants, and certain generalizations about them can be made. For one, they were all men since universities in France remained clerical institutions of the Catholic Church, and as such were bound by the restrictions that kept women outside the ranks of the clergy. The men of the schools also remained detached from the wider intellectual world of the period through their attachment to Latin. In France, the Latinate culture of the university (and of Jesuit and Oratorian *collèges* as well) cut two ways. Not only did it divorce scholastic discourse from the wider vernacular discourse of the literate French public, a divide that was centuries old and common to all European societies of the time, but it also potentially isolated schoolmen from the international discourse of the Republic of Letters, which was increasingly adopting French as its lingua franca. Thus, as a vernacular text, *Les mondes* cut directly across the grain of French scholastic culture in its treatment of natural philosophical topics such as cosmology in the non-scholastic language of French.

Yet Fontenelle was by no means the first to treat natural philosophical questions in this way, and this fact points to the deeper complexities associated with *Les mondes*'s invocation of savant discourse. In France, the great challenge to Latin academic authority in natural philosophy had been offered a half century earlier in the writings of René Descartes. Although he wrote his 1641 *Meditations on First Philosophy* in Latin, he had already published his earlier *Discourse on Method* (1637) in French. There he declared that "if I write in French, which is the language of my country, rather than in Latin, which is that of my teachers, it is because I hope that those who rely purely on their natural intelligence will be better judges of my views than those who believe only what they find in the writings of antiquity."[21] In the *Discourse,* Descartes employed a dichotomy which celebrated the authority of the popular and the vernacular and disparaged that of the scholastic and the Latin. Even in his *Meditations,* written in Latin and dedicated to the "learned Doctors of the Sorbonne," Descartes positioned himself against the grain of the scholastic establishment. For example, the text, which was quickly translated into French with the author's expressed approval and in the same autobiographical idiom pioneered in the *Discourse,* sent a different message. It argued in the language of the French people for a set of principles which challenged fundamental questions of scholastic natural philosophy. It also challenged the institutional power of the French scholastic establishment by challenging the scholastic/Latin monopoly over official learning, including its authority to control and police philosophical discussion itself.[22]

Descartes, therefore, offered not only a new natural philosophy in his works but also a provocative new identity for the natural philosopher. Before the *Discourse,* natural philosophers and university professors were identical. They were supported in their dual identities not only by their technical command of these difficult subjects but also by their command of Latin, the authoritative language of learning, and by their university title, a status attribute that marked them as authoritative speakers on natural philosophical matters. Descartes, by contrast, represented a new species, what the historian Amos Funkenstein has called a "secular theologian."[23] "Secular theologians," or "lay natural philosophers" to coin a synonym, were a new creation of the seventeenth century.[24] They claimed the authority to speak about natural philosophical questions (including those, like mind/body dualism, that contained potent theological implications) without the sanction of traditional scholastic and ecclesiastical authority (the two were conjoined in the early modern French university). Descartes, in fact, was capable of writing in Latin and did so on occasion (his Jesuit education had served him well), but he never held (or sought to hold) a university post. Nevertheless, from his position outside the official scholastic establishment, he claimed the right to speak authoritatively about even the most difficult questions of natural philosophy. In this way, he pioneered a new kind of natural philosophy written by lay people for lay people.

Returning to Fontenelle's appeal to "savants" in *Les mondes,* the contestation surrounding this term in the 1680s, rooted in the continuing tension between scholastic and lay authorities in the adjudication of intellectual matters and in the tension between official, scholastic Latin, and popular, vernacular French, makes interpreting Fontenelle's agendas a tricky matter. The French noun *"docte,"* for example, was typically used in the seventeenth century to isolate—typically in order to disparage—the bookish, Latinate scholars that populated the schools. Yet *"doctes"* could also refer to bookish, Latinate savants outside the schools (i.e. lay natural philosophers) who were attempting to implicate themselves within the scholastic establishment by writing Latin works of their own. They could also be those who shared with Descartes an aversion to scholastic learning and authority, but who nevertheless continued to esteem Latin eloquence, ancient learning, and erudite, text-based scholarship, the founding values of the humanist Republic of Letters.[25]

Fontenelle, moreover, did not target *les doctes* as his audience, but focused instead on savants. To invoke the latter meant to target less the scholastic establishment per se than the contested space that existed between scholastic culture and the wider world of literate lay readers and writers in France. In other words, it was to target the space where learned men, including scholastics and lay natural philosophers, interacted with *gens du monde*—the other partner in Fontenelle's proposed wedding. The *gens du monde* present yet another constituency, for they were members of the literate lay public who possessed no university credentials whatsoever, and who rarely had training in Latin. Yet they were literate and intellectually oriented and also possessed a deep social need to mark themselves as people of quality, a need which led many to pursue serious reading in philosophy and the sciences as part of a program of elite self-fashioning.

Indeed, Descartes and the other lay natural philosophers of the mid-seventeenth century largely secured their institutional authority by developing strong ties to the intellectual world of these non-scholastic *gens du monde.* At the same time, and from the other direction, elites in France likewise found new and powerful possibilities for social maneuver through an attachment to the new culture of lay natural philosophy that Descartes' writings helped to foster. As Erica Harth has explored most fully, Cartesianism found some of its earliest and most ardent adherents in France not within the schools (even if university scholars and clerics such as Régis and Arnauld did become passionate defenders) but within the seventeenth-century intellectual salons of Paris.[26] The participants in these salons were the same literate, non-scholastic, yet intellectually inclined readers to whom Descartes had appealed when he chose to publish in French. They were also, on the whole, elites inclined toward practices of sociability that cut against the grain of the traditional structures of Old Regime society. In particular, since salons were sites of mixed-sex sociability where women often assumed positions of authority, these institutions complicated the gender economy of this otherwise patriarchal society. They especially provided a venue for female writing and intellectual work, a space denied to them by the schools and other intellectual institutions of the period. The powerful union forged in this period between the new philosophy and the new elite, mixed-sex sociability of the *gens du monde* was therefore crucial in solidifying a dual challenge to the existing order.[27]

It was joined by similar unions between Cartesianism and other social formations that, like salons, united the elite, non-scholastic *gens du monde* in new ways. The term "academy," for example, began to acquire a new meaning in precisely this period when, following the example of the Accademia del Cimento in Florence, the word came to be applied not to a school but to a gathering of elite intellectuals devoted to the sociable pursuit of learning.[28] Self-proclaimed academies of this sort sprang up all over Europe in the second third of the seventeenth century, and nowhere was this movement more vigorous than in France.[29] Marin Mersenne was the sponsor of one such academy (although he did not call his gatherings by this name), which played a role in the dissemination of Cartesianism.[30] After Descartes' death, Parisian academies (defined loosely to mean any formalized, intellectual society) grew in number, and many continued to serve as conduits for his ideas. The Academy of M. de Montmor, perhaps the most prestigious of these early Parisian societies, served as the site for the influential Christiaan Huygens's early engagement with Cartesian natural philosophy.[31]

More accessible were the widely popular "Cartesian Wednesdays" of Jacques Rohault where both men and women could learn the foundations of Cartesian science in a setting that was instructive, sociable, and entertaining.[32] Rohault's gatherings were equal parts intellectual society, unaccredited school, and sociable party, and as such they illustrate the blurry line that separated salons from academies in the middle decades of the seventeenth century. Marking this border was a tension between pleasureable diversion, utilitarian instruction, and the disciplined pursuit of knowledge production, a tension that sat at the center of the learned/popular, scholastic/lay, Latin/French divides more generally.

Contemporaries were aware of these tensions. Samuel Sorbière, a regular at the gatherings of M. de Montmor, even went so far as to publish a pamphlet in 1663 decrying the clever wordplay and intellectual gamesmanship that he felt was becoming all too common at this once esteemed scientific institution. Invoking the venerable tradition of the Florentine Accademia and its achievements in the development of the new experimental philosophy (Galileo had been a member), Sorbière called for stricter discipline in academic assemblies as a way of making these gatherings a vehicle for advancing scientific knowledge.[33] In so doing he was drawing a firm line between the salon and the academy as legitimate sites for thinking and talking about nature.

Ruling assumptions about proper discourse and sociability certainly distinguished salons from academies in the seventeenth century, but gender actually constituted the more powerful force shaping this divide. Academies as they developed in the seventeenth century were enclaves of all-male sociability. In this respect, they mirrored the world of the lay natural philosophers, and the international Republic of Letters from which they sprang. They also mirrored the world of the schools against which they fought for institutional authority. Salons, by contrast, were sites of mixed-sex sociability where women enjoyed power in their roles as supervisors of taste, decorum, and cultural authority. Rohault's gatherings were thus an innovation because he offered academic-style sociability to an audience that possessed the mixed-sex character of the salon, as well as a salon-style audience for the masculine discourses and practices of the academies. Molière's satire of the "savant women" who participated in Rohault's gatherings in fact finds its engine precisely in the anomalous character of Rohault's assemblies. The play's humor derived from the unnatural image of women acting as savant participants in serious academic gathering and in the oddity of hearing them speak the alien language of serious philosophy. In this respect, it both illustrated and reinforced this gender economy.

Molière's satire was supported by a wider discourse that coded serious academic inquiry as masculine and playful salon conversation as feminine, thus reinforcing the same gender divide. It was further reinforced when the French crown, inspired in no small measure by Sorbière's critique of academic learning in France, created an official Royal Academy of Sciences in 1666 devoted to the serious pursuit of natural philosophy. This institution collected a number of leading lay natural philosophers into one body, thus constituting a royally sanctioned solidification of the lay challenge to scholastic authority in the sciences. At the same time, the crown also enforced the gender divide by its authorization of an all-male institution to serve the nation through the disciplined pursuit of utilitarian natural inquiry.[34] The overall result was a solidification by 1680 of a clear set of oppositions—pleasure versus utility, diversion versus knowledge, conversation versus demonstration, taste versus truth—as the categories that defined the contested border separating the savant world from the world of the *gens du monde* in seventeenth-century France. Moreover, a powerful discourse had begun to institutionalize a male/female, masculine/feminine divide as an integral component of this framework.

## Remapping French Natural Knowledge: Gender and Philosophy in and around *Les Mondes*

Most scholars today tend to treat *Les mondes* as a natural product of this new environment, one that merely mirrors the emerging gender divide discussed above through its purported translation of male natural philosophy into the female idioms of the salons. I prefer to read it as an interesting and provocative engagement with these developments.

To appreciate Fontenelle's challenge, one must remember that when he wrote his text, these oppositions, while clearly demarcated, were in no way set in stone. The popularity of Rohault's gatherings had indeed caused men such as Sorbière to seek stronger disciplinary divisions between academies and salons, and to encourage more aggressive policing of the border between pleasurable conversation and utilitarian knowledge production. The founding of the Royal Academy of Sciences in 1666 only intensified these efforts. This was especially true when Rohault, who was both a public lecturer and the author of a leading physics textbook that was used at universities throughout Europe well into the eighteenth century, was excluded from the company along with Régis, another popular lecturer in the Rohault mold. In this way, the new royal academy did secure the gendered institutional divide that separated all-male science from mixed-sex sociability.

But at the same time the royal academy also helped to solidify an opposition to its exclusivity since Rohault's union of lay natural philosophy with pleasurable, mixed-sex sociability did not disappear after 1666. If anything it only matured, finding new outlets for its ongoing rivalry with both scholastic and academic learning at the interstices of Louis XIV's system of cultural absolutism. Most interesting in this respect was the birth and remarkable growth of the *Mercure galant* after 1677. The *Mercure* was a monthly periodical that began as the personal newssheet of its editor, Jean Donneau de Visé.[35] Its intended audience was the sociable *gens du monde* who had been attracted to Rohault's assemblies and to Parisian mixed-sex sociability more generally. As the work grew in size and readership, it began to offer a mix of content appropriate to the tastes and sensibilities of these readers. Society news, poems, games, works of fiction, puzzles, essays, and a considerable amount of lay natural philosophy—discussions of Cartesianism were especially common, and several issues in 1697–1698 were devoted to a treatise on algebra[36]—constituted the fare that Donneau de Visé offered to his worldly readers.

The mix personified the fusion of learned culture and elite sociability that had been evolving in France over the previous decades. Yet the *Mercure* added a new dimension to it by constituting itself as a new kind of virtual institution as well. Academies like the one hosted by Mersenne had already established the practice of linking physical sociability with virtual sociability by serving as exchange centers for letters and texts. Other academic institutions both inside and outside of France had done the same, and some, such as the Royal Society of London, formalized this relationship by instituting a periodical that served as the printed, public organ of these correspondence networks. In France, however, neither the royal academies nor their unofficial counterparts had taken a similar step. Donneau de Visé's periodical was thus truly pioneering in its effort to institutionalize a public through the shared reading and discussion of a single, common, periodical text.

Most important was the way that the journal, and the virtual community that it defined, solidified relationships between individuals who were not attached to the official French intellectual establishment, scholastic or academic. In this respect, the *Mercure* constituted an institutional rival, albeit a virtual one, to academic and scholastic authority in France. This rivalry was not always antagonistic, for the different groups that were vying with one another often shared many of the same values and assumptions. Many also participated in more than one site of intellectual exchange, and thus border crossing was the norm rather than the exception. Nevertheless, at a time when institutional borders were being drawn in France, the *Mercure* effectively thwarted any easy hegemony of the academy system over the complex intellectual field of the kingdom.

The power and influence of the *Mercure* and its constituencies was made manifest during arguably the most important intellectual struggle of the period, the "Quarrel Between the Ancients and Moderns." The struggle began within the *Académie française,* the older and more literary brother of the Academy of Sciences, when the academician Charles Perrault read a poem extolling the virtues of contemporary writing when compared with the classical works of antiquity. The poem was greeted by intense disagreements within the academy, and soon the controversy spilled over into the wider public.[37] The *Mercure* was ideally poised to exploit this opportunity. As the debate raged, the periodical played an active role by publishing a number of the key texts, by summarizing and thus fueling many of the key arguments, and by simply encouraging the polemic itself. The journal's sympathies lay squarely with the "Moderns," and even though the battle had begun as an intramural struggle among academicians, and even though its central theater of war remained the Academy itself, the *Mercure* helped to foster the powerful perception that it was instead an institutional struggle that pitted the "old guard" of the French academic establishment against the young turks of the salons and *mondain* society. Indeed, in a provocative and persuasive analysis of "the Quarrel," the literary critic Joan

DeJean has likened it to the American cultural wars of the last two decades, arguing that the "Ancients" were akin to American conservatives fighting to defend the traditional canon while the "Moderns" played the role of seventeenth-century "postmodernists," "feminists," and "multiculturalists" determined to expand official culture in the name of new political and cultural priorities.[38]

However far one wants to take DeJean's analogy, it is clear that an institutional struggle over the control of knowledge was at the center of the Quarrel. It is also clear that gender played a key role in it. Focusing on fiction writing, DeJean shows how the controversy was centered on the relative merits of "masculine" and "feminine" forms of writing (i.e. epic, tragedy, and heroic prose versus lyric and the novel respectively). She also shows how the presence and increasing influence of female literary production, especially in the form of the new female-authored novels such as Lafayette's *La Princesse de Clèves,* helped to turn these struggles into real political battles over female participation in the production of culture. In this respect, the Quarrel mirrored in a more overtly polarized and polemical way the struggles over female participation in the production of natural philosophy that the new Cartesian lay natural philosophy had occasioned. Indeed, since DeJean believes that the Quarrel achieved its peak of intensity only when its many polemics were distilled into one—the proper place of women in the sphere of literary production—it very well could be argued that the Quarrel exposed the gender tensions implicit in the changing institutional environment of French intellectual life as a whole and made them visible for self-conscious debate and critique.[39]

The two struggles certainly mirrored each other ideologically if not in terms of polemical heat since each produced a recognizable feminist discourse that argued for the positive evaluation of female knowers and for the inclusion of women in the field of knowledge production. Within the struggle between academies and salons over lay natural philosophy, the key text was François Poullain de la Barre's *De l'Egalité des deux sexes,* first published in 1673. Explicitly invoking Descartes' dualism between mind and body, Poullain de la Barre argued that in the realm that mattered, the mind, no difference between men and women existed. "The mind has no sex," he famously declared, and thus there was no reason that women could not equal men in all branches of learning so long as society gave them the education and the support necessary to develop their talents.[40] Within the Quarrel, it was not an "equality feminism" but a "difference feminism" that dominated. Perrault, the poet who sparked the Quarrel itself, expressed the Modern position with greatest clarity when he argued in his *Parallèle des Anciens et Modernes* and then again in his *Apologie des femmes* that "women's judgment" should serve as the model for authority, taste, and genius.[41] It was not necessary that men become women, Perrault averred, but it was necessary that they learn "to think, to judge, and to reason as a woman" if they were going to achieve the excellence to which they aspired.[42] This formulation clearly challenged the authority of all-male institutions such as the academies, but rather than calling for female inclusion in such bodies as the antidote to masculine domination, it validated mixed-sex sociability and defended practices of cultural production where men and women formulated knowledge together through a shared and intersubjective exchange of difference.

Perrault's views were given ample reinforcement in the pages of the *Mercure,* where they joined seamlessly with the less boisterous philosophical discussions of the period and with a plethora of other male-female intellectual exchanges. In this way, the journal helped to crystallize a particular (if internally contested) modern identity within its virtual community founded upon a loose and not always coherent set of ideas about sociability, aesthetics, philosophy, language, intellectual practice, and gender. This complex provided the motivation and the architecture for Fontenelle's *Les mondes,* and the source for it was none other than the *Mercure* itself. As a young provincial fleeing his parents' careerist aspirations for him, Fontenelle found his first intellectual home as a journalist working for Donneau de Visé. He quickly established a name for himself by publishing in the *Mercure* and elsewhere, and he reinforced his rapidly growing reputation through his easy navigation of Parisian society.[43] When the Quarrel broke out, Fontenelle chose sides easily since Perrault and the other Moderns were at the center of his coterie. He soon joined the debate himself, publishing his own polemical pieces in support of Perrault's position and employing his increasing cultural authority to swing the public toward the Moderns' point of view.[44] *Les mondes* was in fact published only a few months before Perrault's poem triggered the Quarrel itself, and consequently the work speaks to and from this controversy and to Fontenelle's position within it as well as to the wider conflicts that he and the other antagonists activated.

Yet more than the Quarrel it is Fontenelle's appointment as Perpetual Secretary of the Royal Academy of Sciences in 1697 that looms over interpretations of his most widely read text. By entering the all-male world of the Academy and by becoming the public spokesman for its disciplined codes of gentlemanly sociability and utilitarian knowledge production, Fontenelle is said to have left behind the world of the Moderns and its mixed-sex ethic of pleasurable natural inquiry. Read through this understanding of the 1697 appointment, *Les mondes* is also interpreted as a foreshadowing of this new role even though its publication preceded Fontenelle's entry into the academy by over a decade.[45]

I am not inclined to argue with this representation of the gender significance of Fontenelle's appointment as Academy secretary, even though I do think that the historical nuances of the institutional situation deserve more attention than has been given to them. What I do contest is the argument that *Les mondes* replicates in textual form the relationship between all-male academic science and feminized sociability implicit in Fontenelle's new role as Academy secretary. The book suggested a very different relationship between men and women in the production of knowledge than the one institutionalized through Fontenelle at the Royal Academy, and I would suggest that Fontenelle's invitation to join the Academy (based, it should be remembered, on almost no other scientific credentials than his authorship of *Les mondes*) may have been less a "co-optation" of the "feminized public" by the all-male academy than a temporary concession to it, a concession that acknowledged the power and influence of the rival institutional configuration exemplified in *Les mondes*. The concession probably did not mark anything more than a temporary diversion from the otherwise triumphant emergence of masculine scientific and academic authority. Yet it nevertheless seems worthwhile to acknowledge and explore this historical eccentricity and to emphasize its significance as a lost historical alternative. A more careful and precisely situated reading of *Les mondes* opens the door to this different understanding of the book's significance.

## *Les Mondes* and Fontenelle's Dialogic Natural Philosophy

The most important feature of *Les mondes* from the point of view of its wider philosophical significance is the fact that Fontenelle chose to present his ideas in the form of a dialogue. This choice was not overdetermined by the existing intellectual and genre assumptions of the 1680s. Descartes had laid the foundations for the lay natural philosophy which Fontenelle sought to continue not by writing dialogues but by developing an innovative form of philosophical autobiography as his preferred method of presentation.[46] Viewed institutionally, Descartes' autobiographical approach, which featured an intense emphasis upon his own subjectivity and an effort to reground philosophical authority in the cogitations of a single, solitary subject, constituted yet another challenge to the prevailing scholastic authorities.

Yet despite its private challenge to the philosophical authority of public, scholastic disputation, Descartes' autobiographical approach to philosophy remained anchored in the method of dialectic, especially in its emphasis upon rational demonstration to a universal public. In this respect, Fontenelle's dialogic philosophy was anything but Cartesian. Instead, *Les mondes* drew its inspiration from other forms of early modern European writing that assumed a different notion of intersubjective, non-dialectical knowledge production. The court literature inspired by Castiglione's influential *Book of the Courtier* was one source for *Les mondes*. Castiglione's work used the dialogue form to stage a sociable conversation among a mixed-sex gathering of elites. In this way, it became a model for countless other works that similarly wanted to present knowledge as a consequence of civil sociability.

The contrast between lay natural philosophy à la Descartes and courtly philosophy à la Castiglione was great, and the tensions between them, which were acknowledged at the time, were the textual counterparts to the institutional divisions between academies and salons discussed above. But as was the case with these wider institutional tensions, the divide between these genres was not absolute. Galileo, for example, chose to present a significant portion of his natural philosophy in the form of dialogues, and as the historian Mario Biagioli has suggested, Galileo's connections to the court culture of the Medici family played no small part in this decision.[47] Galileo's dialogues, however, were staged as all-male exchanges, and their setting suggests scholasticism, or perhaps an all-male academy, rather than courtly society. Yet his stylistic panache, which has earned him an esteemed place in the pantheon of Italian letters, brings him back closer to Castiglione, as does the intellectual style of the dialogues which resembles the witty repartee of Montmor's academy more than a Cartesian demonstration or a scholastic disputation.[48]

In France, these same genre and language tensions emerged in tandem with the institutional struggles associated with lay natural philosophy. Alongside Descartes' genre innovations, other writers adopted the dialogue form as their preferred vehicle for philosophical exposition. Some, such as Malebranche in his *Entretiens sur la metaphysique et sur la religion,* held tightly to the model of the dialectical disputation even if the decision to eschew Latin in texts such as these marked them as lay alternatives to scholastic discourse.[49] Others, such as the Jesuit Father Dominique Bouhours in his *Entretiens d'Ariste et d'Eugene,* remained more closely tied to Castiglione.[50] Still others fell somewhere in between. Rohault, for example, authored both a philosophical treatise and a philosophical dialogue in support of his Cartesian views, and the differences between the two are interesting. The treatise is structured as a book-length demonstration, with the early pages devoted to basic principles and the latter parts to rational demonstrations. Not surprisingly, it became the standard text in European universities once Cartesian science became widely accepted.[51] Rohault's dialogue, by contrast, offers an exchange between a man and a woman, and it proceeds through questions and long explanatory answers without ever adopting the demonstrative rigor of the treatise. In this way, the dialogue reflects Rohault's "Cartesian Wednesdays," especially in the gender arrangement.[52]

Fontenelle's *Les mondes* represented yet another approach, and it is precisely in his genre innovations that the significance of Fontenelle's philosophy is revealed. The author's self-professed desire to "treat philosophy in an unphilosophical manner" was realized through a genre hybridization that fused the different approaches to lay natural philosophy then available in France. On the one hand, *Les mondes* strongly embraced Castiglione's notion of courtly philosophy with all of its pleasurable discursive play and its mixed-sex ethic of intersubjective, interactive knowledge production. On the other hand, the text also attempted to hold on to the more demonstrative approach that characterized the dialogues of Galileo and Rohault.

A comparison of Rohault's and Fontenelle's *Entretiens* reveals the innovative nature of the latter's approach. While both stage a dialogue between a man and a woman, Rohault's is much more hierarchical, with the male narrator clearly assuming the role of teacher and demonstrator and the female interlocutor assuming the role of dutiful pupil. The relationship between Fontenelle's narrator and his Marquise is much more complex. The intellectual progression of the two dialogues is also quite different, with Rohault's moving in a linear fashion from ignorance to knowledge while Fontenelle's moves back and forth in a nonlinear fashion between moments of inquiry and insight. Similarly, whereas Rohault's dialogue situates the woman as a passive recipient of the male knowledge provided by her instructor, Fontenelle's text, despite the assertions to the contrary made by such scholars as Harth and Terrall, stages a much more intersubjective encounter between two different but equally important interlocutors. The result is a text that models an inescapably dialogic approach to natural inquiry and assumes a powerful role for both men and women in the enterprise of natural philosophy.

To examine the details of Fontenelle's philosophy in more detail, consider first his statements about the presence of the Marquise in the text. "I've placed a women in these Conversations who is being instructed," he explains in this "Preface." "I thought this fiction would serve to make the work more enticing, and to encourage women through the example of a woman who, having nothing of an extraordinary character, without ever exceeding the limitations of a person who has no knowledge of science, never fails to understand what is said to her."[53] These statements connect *Les mondes* to the feminist pedagogical project outlined by Poullain de la Barre. Drawing out the point, Fontenelle concludes: "Why would any woman accept inferiority to this imaginary Marquise, who only conceives of those things of which she can't help but conceive?"[54] Rohault's *Entretiens* likewise positioned itself in terms of a pedagogy of female self-improvement, yet for Fontenelle this paternalist agenda is not the impetus of the book. "To be honest," he concedes, "the Marquise applies herself a bit," and then opening a very different pathway for female learning and for natural inquiry more generally than that offered by Rohault, Fontenelle added: "but what does applying oneself mean in this context?"

> It's not necessary to penetrate by means of concentrated thought something either obscure in itself or obscurely explained; it's merely required that one read and at the same time for one to form a clear idea of what one is reading. I only ask of ladies for this whole system of philosophy the same amount of concentration that must be given to *The Princess of Cleves* in order to follow the plot closely and understand all its beauty. Its true that the ideas of this book are less familiar to most women than those of *The Princess of Cleves,* but they're no more obscure; one cannot read them more than twice at the very most without grasping them very accurately.[55]

Such an approach to philosophy could not have been more different than the invitation to deep, probing self-examination that Descartes offered in his *Meditations*. It was also very different from the pedagogical agendas of Rohault, who wanted to bring women into the fold of male natural philosophy by assuming their intrinsic equality and by then teaching them the discourses and practices of the world from which they had been excluded. Fontenelle, by contrast, fused the male world of academic science with the mixed-sex world of the salons, creating a new hybrid. It is a world where proper philosophical practice and exposition are in no essential way different from proper literary practice and exposition. Implicit in this fusion is the further claim that the all-male world of academic science in fact departs from the true path of knowledge when it claims the discipline of "concentrated thought" as epistemologically essential or "obscure reasonings" as a necessary and intrinsic part of scientific practice itself. Instead, like Perrault, who encouraged men to "reason like women," Fontenelle encouraged both men and women to embrace an approach to philosophy that is as unlabored and pleasurable as it is clear and persuasive.

Fontenelle in fact connected this view to the pursuit of natural inquiry as a whole in his subsequent paragraph. Insisting that his is not a work of fiction and that he had no intention of "creating a make-believe system," he nevertheless pointed out the natural harmony that exists between truth and pleasure when both are pursued appropriately. "Fortunately it happens . . . that the ideas of physics are pleasing in themselves and, at the same time, that they are satisfying to the mind. They provide a spectacle for the imagination which pleases it as much as if they had been made expressly for that purpose."[56]

Framed in this way, physics, pleasure, reason, and the imagination are not at odds with one another but are mutually reinforcing. This outlook is key to Fontenelle's philosophy of natural inquiry as it is revealed in *Les mondes*. Articulating this view at the very beginning of the text, Fontenelle's narrator announced his beliefs about the plurality of worlds in an interesting and revealing way. "I wouldn't swear that it's true," he declares, "but I think it is so because it pleases me to think so. The idea sticks in my mind in a most delightful way. As I see it, this pleasure is an integral part of truth itself."[57] The same theme is repeated at other points in the dialogue, and taken as a whole these statements amount to a complex philosophy of knowledge, a philosophy articulated in *Les mondes*.

For Fontenelle, knowledge about the essential constitution of nature was impossible. Indeed, essences do not even exist, for as he stated elsewhere: "God has no universal ideas; his infinite understanding embraces all particulars and he has no need to abstract or abbreviate. . . . Consequently a universal can represent nothing real and in fact represents nothing at all."[58] Furthermore, since sensate knowledge of nature's particulars is the only knowledge humans can have—Fontenelle was at core an empiricist as well—natural knowledge is restricted to inferences drawn from limited sensate experience. Here Descartes' attempt to make dialectical reason the Archimedean center of a new demonstrative science was shown to be misguided. Instead, the imagination rather than reason played the crucial role since natural philosophy ultimately amounts to imaginative picture making, pictures that are then tested against a sensate understanding of the world.

In *Les mondes,* oriented as it was toward the project of making natural philosophy compatible with urbane sociability, the suggestion was made that the most credible pictures are also the most pleasing to the mind. As the narrator stated, "pleasure is an integral part of truth itself." But this aesthetic criteria for truth is more than just a literary flourish. It represents a philosophically rigorous position as well. For in the dialogue, Fontenelle made it clear that truth is not arbitrary or subjective. "It would seem that your philosophy is a kind of auction," the Marquise states early in the text, "where those who offer to do these things at the least expense triumph over others." "It is true," the narrator replied, "and it is only by that means that one can catch the plan on which Nature has made her works. She's extraordinarily frugal. Anything that she can do in a way which will cost a little less, even the least bit less, be sure she'll do it only that way. This frugality, nevertheless, is quite in accord with an astonishing magnificence which shines in all she does. The magnificence is in the design, and the frugality is in the execution."[59]

In this interchange, Fontenelle articulated the central epistemological conviction of *Les mondes.*[60] Humans are restricted to the pictures of the cosmos that our imagination creates out of the limited sensate experience we have. Any scientifically sound picture must account for the observed phenomena, and thus science centers on separating the good pictures from the bad. Yet no empirically sound picture is more necessarily true than any another. In fact, given several equally sound empirical accounts there is no necessary reason for preferring one over another. Nature, however, does obey a strict law of economy whereby whatever is simplest is always preferred. Consequently, given a set of possible scientific pictures of the universe, Fontenelle ultimately argued that the most elegantly simple and economical was always to be preferred. Importantly, it was not preferable because some necessary metaphysics made it the one and only true system of the universe; rather, it was preferable because its simplicity and economy suggested that no better or more natural picture was possible.

Turning this position around, a powerful scientific epistemology follows from it, one rooted in a kind of empirical nominalism. Confronted with a world of infinite particulars, human beings develop a scientific understanding of nature by creating reasonable pictures to explain what they see. Many different pictures are possible, and no manner of logical deduction or philosophical reasoning can ever show us which picture is true and which is a mere illusion. Indeed, as Fontenelle stated elsewhere: "What we commonly call logic . . . is vain and useless research, . . . of importance only to its author who has not completed it without hard work and a great deal of cleverness."[61] Our mental pictures of nature, moreover, lacking any natural necessity, also never succeed in collapsing the gap between human imagination and the true character of nature. In fact, since humans are incapable of acquiring the complete knowledge of all particulars that God possesses, the science/nature gap can never be collapsed empirically either. Yet we are able to differentiate between good and bad scientific pictures because given any two empirically sound accounts of a natural phenomena, we can rest assured that the simpler one is closer to nature's true behavior. *Les mondes,* in its systematic exploration of pleasurable picture making and picture testing, articulates this philosophy throughout. The same message was reiterated throughout much of Fontenelle's other work.

Crucially for the argument here, this philosophy also offers a very different understanding of the gender economy appropriate to natural scientific inquiry. Scientific picture making properly understood requires attention to both empirical truth and aesthetic beauty. It also depends upon an epistemology that unites a concern for rational clarity with an equally powerful interest in pleasure and the satisfaction of the imagination. Furthermore, the picture testing that is required before any idea is accepted requires an interactive environment and a commitment to free discursive play since reason and the imagination must work together if science of this sort is to succeed. The notion of critical inquiry and philosophic judgment operative here is thus deeply connected to Perrault's notion of male-female interaction in the production of aesthetic truth. Indeed, since Fontenelle's dialogic natural philosophy placed an epistemological premium on the aesthetic, and since it also considered pleasure and imaginative satisfaction as central values, it is appropriate to consider Fontenelle's natural philosophy and Perrault's aesthetic philosophy as but two sides of one coin. From this perspective, it is also appropriate to consider each as defenders of an intersubjective understanding of truth that assumes mixed-sex sociability and cross-gender intellectual exchange as core knowledge producing practices.

Fontenelle's *Les mondes* reads more fluidly when this intersubjective and inter-gender conception of knowledge production is used to frame an understanding of the dialogue. Far from a scholastic disputation, or a public scientific demonstration, the dialogue best resembles a lively give-and-take between two assertive and equally talented interlocutors. True, the dialogue opens with the Marquise seeking to draw knowledge from her interlocutor by inciting the male narrator to speak about what he knows. It similarly concludes with the final proclamation of the Marquise that "I am now *savante!*" In this way the text, on a superficial level at least, demonstrated male knowledge passing to a receptive female audience through sociable conversation.

Yet the dialogue itself defies such simple characterizations. The narrator, for example, is anything but a dogmatic knower. From the very outset his views are offered as "peculiar" and as "his own strange conceptions." "You are mad," the Marquise asserts at one juncture, to which the narrator replies: "Who's arguing?"[62] Moreover, these statements are neither disingenuous nor mere rhetorical flourishes, for the narrator repeatedly insists that he cannot be relied upon to reveal certain truth. "May these systems not, despite this similarity, differ in a thousand ways?" the Marquise asks near the end of the text. "Definitely," the narrator responds, ". . . but what do I know?" Continuing with a brief summary of the possibilities about what might be true, the narrator concludes, "what do you want from me? That's enough for a man who's never left this vortex."[63] Fontenelle's probabilist epistemology, which denies the possibility of certain truths about nature, is at work here, but so is the centrality of intersubjective dialogue in the production of knowledge. For if no single demonstration can ever reveal truth, the only thing that can reveal it, or at least allow humans to approach it, is the mutual exchange of ideas between two equal and complementary interlocutors. The message is driven home in the final passage of the dialogue, when after assenting to the Marquise's declaration that she is now a *savante,* the narrator continues: "Yes, you are well enough, and you've the advantage of believing nothing at all of what I've told you whenever you choose."[64]

For her part, the Marquise plays her role in this exchange with brilliance and assertiveness. At the outset, she forces the narrator into the dialogue itself by refusing to let him invoke the gender divide between them and the discourses of masculine and feminine sociability that reinforced it. "If your idea is pleasing, then share it with me," the Marquise requests. But, the narrator replies, "this won't produce enjoyment such as one would find in a Molière comedy; it's enjoyment that involves our reasoning powers. It only delights the mind." "What?" she fires back. "Do you think I am incapable of enjoying intellectual pleasures? I'll show you otherwise right now. Tell me about your stars!" Sheepishly the narrator tries to resist one last time, invoking the shame of having it known that he "spent an evening with a most beautiful woman speaking about philosophy." The Marquise will have none of it. The narrator begins to reveal his thoughts and the dialogue follows in train.[65]

Furthermore, having established her assertiveness at the outset, the Marquise continues to drive the dialogue itself. She never shifts roles, nor offers any knowledge herself, but she also never lets the narrator stray from her agenda. At times, the Marquise demands clarifications, forcing the narrator to become more explicit. At other times, she admits consent, thus allowing the narrator to move to other topics, but usually based on her question or suggestion. On still other occasions she challenges the narrator's account, such as when she accuses Copernicus of calumny for humiliating human beings by de-centering them from the cosmos. "We should never have accepted his system," she opines.[66] On these occasions of contestation, which are many, the narrator never adopts a disciplinary mode, nor attempts to illuminate the Marquise's ignorance through a didactic presentation of knowledge. Instead, disagreement is presented as an opportunity for further dialogue where the narrator and the Marquise together explore ideas in a collaborative search after consent. The first *soir* in fact is almost entirely occupied with the implications of Copernican de-centering, and while the dialogue ends with the narrator telling us that the couple "resolved in the end to hold to the system of Copernicus" because it is "more uniform and enticing and free of prejudice" and because "its simplicity is persuasive and its boldness pleasing," we learn about this resolution not through the dialogue itself but through an after-the-fact report by the narrator during his walk back home from the garden. The dialogue proper ends with the Marquise calling a halt to the discussion because of the boredom it is generating, and she sets the agenda for the next day by asserting that "tomorrow we will come back here, you with your systems and me with my ignorance."[67]

Exchanges such as these are the norm in *Les mondes,* and through them Fontenelle articulates the complicated, intersubjective, and essentially mixed-sex method by which natural knowledge should be produced. Rooted in the polite norms of salon sociability, the method is conversational, playful, and diverting. It is also self-consciously opposed to scholastic and academic pedantry, and assertive in its belief that the true picture of nature must be as pleasing to the imagination as it is to the mind and as pleasurable and beautiful as it is rational and explanatory. Its goals, however, are truth about nature in the manner of both philosophy and literature. Codes of sociable decorum also inform the disciplinary structures of the dialogue and the natural philosophy it articulates. Disagreements, while commonplace, are never allowed to turn into outright arguments or debates, nor are they resolved by one person assuming the role of knower while the other assumes the role of passive pupil and listener. Instead, possibilities are explored and tested in a spirit of mutual exchange and artful play. In the end, either consent is reached or an agreement to defer judgment until later is made, even if it means leaving a question unresolved and in need of further investigation.

In this way, *Les mondes* models a fully dialogic approach to natural inquiry, one that assumes an essential place for both men and women in the production of knowledge and one that marks sites of mixed-sex conversation, diversion, and pleasure as the supreme venues for natural philosophical inquiry. It is, in short, the dialogic equivalent of Descartes' dialectical *Discourse on the Method of Rightly Conducting Reason for Seeking Truth in the Sciences,* but one in which reason and taste and truth and pleasure admit no distinction. In fact, were we to give *Les mondes* Descartes' title, it could appropriately be called *Discourse on the Method of Rightly Conducting Reason and Taste for Seeking Truth and Pleasure in the Sciences.* By integrating reason and taste and truth and pleasure in this way, Fontenelle offered a manual for how natural philosophy and the mixed-sex sociability of seventeenth-century French society were not only compatible but mutually reinforcing. In doing so, he also left us a monument to a lost conception of natural knowledge, one where men and women were not confined to exclusionary spheres but joined in spirited co-production.

## Notes

1. A classic summary account of "the Newtonian Synthesis" is Alexandre Koyré, "The Significance of the Newtonian Synthesis," in *Newtonian Studies* (Chicago: University of Chicago Press, 1965), 3–24.

2. A. R. Hall, *The Scientific Revolution 1500–1800: The Formation of the Modern Scientific Attitude* (Boston: Beacon Press, 1962), 244.

3. Ibid., 244–45

4. By "funnel narrative of the Newtonian Revolution," I mean the way that the classic historiography of early modern mathematics and physics, by focusing so singularly on the "triumphant" Newtonian achievement in the *Principia,* inevitably transforms all of the history preceding and succeeding it into a story of how Newton's triumphal work was produced and then disseminated.

5. Important works in this regard are Richard S. Westfall, *Never At Rest: A Biography of Isaac Newton* (Cambridge: Cambridge University Press, 1980); Robert E. Schofield, *Mechanism and Materialism: British Natural Philosophy in an Age of Reason* (Princeton: Princeton University Press, 1970); B. J. T Dobbs, *The Janus Face of Genius* (New York: Cambridge University Press, 1992); Larry Stewart, *The Rise of Public Science: Rhetoric, Technology, and Natural Philosophy in Newtonian Britain, 1660–1750* (Cambridge: Cambridge University Press, 1992); Margaret C. Jacob, *The Newtonians and the English Revolution* (Ithaca: Cornell University Press, 1976); Margaret C. Jacob, *The Radical Enlightenment: Pantheists, Freemasons, and Republicans* (London: Allen and Unwin, 1981); Michel Blay, *La naissance de la mécanique analytique. La science du mouvement au tournant des XVIIe et XVIIIe siècles* (Paris: Presses Universitaires de France, 1992); and Niccolò Guiciardini, *Reading the* Principia: *The Debate on Newton's Mathematical Methods for Natural Philosophy from 1687–1736* (Cambridge: Cambridge University Press, 1999).

6. Londa L. Schiebinger, *The Mind Has No Sex? Women in the Origins of Modern Science* (Cambridge, MA: Harvard University Press, 1991); Erica Harth, *Cartesian Women: Versions and Subversions of Rational Discourse in the Old Regime* (Ithaca: Cornell University Press, 1992); Mary Terrall, "Gendered Spaces, Gendered Audiences: Inside and Outside the Paris Academy of Sciences," *Configurations* 2 (1995), 207–32; Mary Terrall, "Metaphysics, Mathematics, and the Gendering of Science in Eighteenth-Century France," in *The Sciences in Enlightened Europe,* ed. William Clark, Jan Golinski, and Simon Schaffer (Chicago: University of Chicago Press, 1999), 246–72.

7. Nina Rattner Gelbart, preface to *Conversations on the Plurality of Worlds,* by Bernard le Bover de Fontenelle, trans. H. A. Hargreaves (Berkeley: University of California Press, 1990), xviii–xix. This edition, from which I take several translations, will hereafter be referred to as Fontenelle, *Conversations.*

8. Emma Spary, "The 'Nature' of Enlightenment," in Clark, Golinski, and Schaffer, *The Sciences in Enlightened Europe,* 272–304; Emma Spary, *Utopia's Garden: French Natural History From Old Regime to Revolution* (Chicago: University of Chicago Press, 2000); Jessica Riskin, *Science in the Age of Sensibility: The Sentimental Empiricists of the French Enlightenment* (Chicago: University of Chicago Press, 2002).

9. Riskin, *Science in the Age of Sensibility,* 7.

10. The best modern, critical edition of this work is found in Alain Niderst, ed., *Oeuvres complètes de Fontenelle,* 9 vols. (Paris: Fayard, 1991–2002), 2:7–130. Hereafter I will refer to this edition as *OF.*

11. It should be emphasized that I am using the term "philosophy" in a very historical sense, and that its meaning as I use it may conflict with our modern notion of the same concept. For an eighteenth-century definition of "philosophy," see the article *"Philosophe [Philosopher]"* in Diderot and d'Alembert's *L'Encyclopédie, ou Dictionnaire raisonné des sciences et des arts.* This article has recently been translated into English by Dena Goodman and is available at http://www.hti.umich.edu/cgi/t/text/text-idx?c=did;cc=did;rgn=main;view=text;idno=did1111.0001.001;sid=105afa856ad51f861

b57b208b7f2ad16. Robert Darnton also discusses the historical eccentricities of the term "philosophy" in Old Regime France in *The Literary Underground of the Old Regime* (Cambridge, MA: Harvard University Press, 1982).

12. Leonhard Euler's *Lettres à une Princesse d'Allemagne,* first published in St. Petersburg in 1768 and reissued in 1772, explicitly invokes Fontenelle's *Les mondes* as its model, as did numerous other works.

13. The classic definition of the "Two Cultures" is found in C. P. Snow, *The Two Cultures* (Cambridge: Cambridge University Press, 1998).

14. A second edition of the text, first published in 1687, included a sixth evening, but this addition, while adding substance to the discussion, in no way changed the overall logic of the text. The author made no further changes to the text, and thus after 1687 *Les mondes* was complete, appearing ever after in either its original five *soirs* format or in the longer six *soirs* version. The English translation, Fontenelle, *Conversations,* is based on the 1686 edition and has only five *soirs*. The edition in *OF* is based on the 1687 edition and has six *soirs*.

15. See Michael J. Crowe, *The Extraterrestrial Life Debate, 1750–1900: The Idea of a Plurality of Worlds from Kant to Lowell* (Cambridge: Cambridge University Press, 1986).

16. *OF,* 2:9–10.

17. L. W. B Brockliss, *French Higher Education in the Seventeenth and Eighteenth Centuries: A Cultural History* (Oxford: Oxford University Press, 1987).

18. François de Danville, "Enseignement scientifiques dans les collèges des Jésuites," in *Enseignement et diffusion des sciences en France au dix-huitième siècle,* ed. René Taton (Paris: Hermann, 1986), 27–64. For a discussion of the place of Jesuit education in the development of seventeenth-century science, see Peter Dear, *Discipline and Experience: The Mathematical Way in the Scientific Revolution* (Chicago: University of Chicago Press, 1995).

19. Alain Niderst, *Fontenelle* (Paris: Plon, 1991), 14.

20. Pierre Costabel, "L'Oratoire de France et ses collèges," in Taton, *Enseignement,* 67–92.

21. Laurence J. Lafleur, trans., *Descartes: Philosophical Essays* (London: Macmillan, 1964), 67, 56.

22. An interesting recent discussion of Descartes' place in seventeenth-century French cultural life is found in Stéphane Van Damme, *Descartes* (Paris: Presses de Sciences Po, 2002).

23. Amos Funkenstein, *Theology and the Scientific Imagination from the Middle Ages to the Seventeenth Century* (Princeton: Princeton University Press, 1986).

24. "Secular theologian" is a synonym for "lay natural philosopher," because the term "natural philosophy" connotes a union of what we today call "science" and "theology." Prior to the disciplinary division of each, to pronounce on natural philosophical matters was at the same time to pronounce, at least implicitly, on theological matters as well. Thus the lay challenge to clerical authority which Funkenstein explores was rooted in theology and natural philosophy simultaneously. Not surprisingly lay figures such as Descartes, Leibniz, Galileo, and Newton figure prominently in his discussion.

25. The prominence of the latter in Descartes' time is emphasized in Peter N. Miller, *Peiresc's Europe: Learning and Virtue in the Seventeenth Century* (New Haven: Yale university Press, 2000). See also Anthony Grafton, *Defenders of the Text: The Traditions of Scholarship in An Age of Science, 1450–1800* (Cambridge, MA: Harvard University Press, 1991).

26. Harth, *Cartesian Women,* esp. chap. 2.

27. See Carolyn Lougee, *Le paradis des femmes: Women, Salons, and Social Stratification in Seventeenth-Century France* (Princeton: Princeton University Press, 1976); and on the eighteenth-century salons, see Dena Goodman, *The Republic of Letters: A Cultural History of the French Enlightenment* (Ithaca: Cornell University Press, 1994).

28. See Mario Biagioli, "Le Prince et les Savants: La Civilité scientifique au 17e siècle," *Annales: Histoire, Sciences Sociales* 50 (1995), 1417–53; Harcourt Brown, *Scientific Organizations in Seventeenth-Century France (1620–1680)* (New York: Russell and Russell, 1967); and Martha Ornstein, *The Role of Scientific Societies in the Seventeenth Century* (Chicago: University of Chicago Press, 1938).

29. Robert A. Schneider, "Openness and Discretion: Cultural Boundaries in the Age of Richelieu," unpublished manuscript. Robert Schneider has identified as many as thirty such institutions in Paris alone during the reign of Louis XIII (1603–1643), and not surprisingly many became sites of the new Cartesian philosophy.

30. Brown, *Scientific Organizations,* chaps. 2–3; Peter Dear, *Mersenne and the Learning of the Schools* (Ithaca: Cornell University Press, 1988).

31. The best account of the Montmor Academy is found in Brown, *Scientific Organizations,* chaps. 4–6. See also Tad Schmaltz, *Radical Cartesianism: The French Reception of Descartes* (Cambridge: Cambridge University Press, 2002).

32. See Harth, *Cartesian Women,* 15–17, 87, 141; Geoffrey Sutton, *Science for a Polite Society: Gender, Culture, and the Demonstration of Enlightenment* (Boulder, CO: Westview, 1995), chaps. 4–5; and Michael R. Lynn, "Enlightenment in the Republic of Science: The Popularization of Natural Philosophy in Eighteenth-Century Paris" (PhD diss., University of Wisconsin, 1997), chap. 1.

33. *Discours prononcé le 3 d'Avril 1663 à l'ouverture de l'Académie des Physiciens, qui s'assemblent tous les Mardis chez Monsieur de Montmor.* BN Mss. f. fr. Cinq Cents de Colbert, 485, folio 441–44.

34. On the founding of the French Royal Academy of Sciences, see John M. Hirschfield, *The Académie Royale des Sciences 1666–1683* (New York: Arno Press, 1981); and Roger Hahn, *The Anatomy of a Scientific Institution: The Paris Academy of Sciences, 1666–1803* (Berkeley: University of California Press, 1971). On the culture of the early academy, see Alice Stroup, *A Company of Scientists: Botany, Patronage, and Community at the Seventeenth-Century Parisian Royal Academy of Sciences* (Berkeley: University of California Press, 1990).

35. Monique Vincent, *Donneau de Visé et le Mercure galant* (Lille, 1987).

36. "Ouvrage concernant l'Algebre," *Mercure galant* (April, 1697), 42–83.

37. Charles Perrault, *Le Siècle de Louis le Grand* (Paris: Jean-Baptiste Coignard, 1687).

38. Joan DeJean, *Ancients Against Moderns: Culture Wars and the Making of a Fin de Siècle* (Chicago: University of Chicago Press, 1997).

39. Ibid., esp. 66–77.

40. François Poullain de la Barre, *De l'égalité des deux sexes* (Paris: DuPuis, 1673). For an English translation of this text, see Gerald M. Maclean, ed., *The Woman As Good As The Man, Or the Equality of Both Sexes* (Detroit: Wayne State University Press, 1988). See also Harth, *Cartesian Women,* 135–39.

41. Charles Perrault, *Parallèle des Anciens et des Modernes, en ce qui regarde les arts et les sciences,* 4 vols. (Paris: Jean-Baptiste Coignard, 1688–1697); Charles Perrault, *Apologie des femmes* (Paris: Veuve de Jean-Baptiste Coignard, 1694).

42. Cited in DeJean, *Ancients Against Moderns,* 67.

43. Niderst, *Fontenelle,* chap. 2.

44. See especially "Digression sur les anciens et modernes," in *OF,* 2:411–31.

45. For this analysis, see Harth, *Cartesian Women,* 123–24; and Terrall, "Gendered Spaces," 212, 217.

46. Interestingly, Descartes did compose a philosophical dialogue, but he left it unfinished and unpublished. *The Search After Truth by the Light of Nature,* in *Philosophical Essays and Correspondence,* ed. Roger Ariew (Indianapolis, IN: Hackett, 2000), 315–23.

47. Mario Biagioli, *Galileo, Courtier: The Practice of Science in the Culture of Absolutism* (Chicago: University of Chicago Press, 1993).

48. Galileo, *Dialogue Concerning the Two Chief World Systems,* trans. Stillman Drake (Berkeley: University of California Press, 1967).

49. Nicolas Malebranche, *Entretiens sur la metaphysique et sur la religion* (Rotterdam: Reinier Leers, 1688).

50. Dominique Bohours, *Les entretiens d'Ariste et d'Eugene* (Paris: s.n., 1671).

51. Jacques Rohault, *Traité de physique,* 2 vols. (Amsterdam: Jacques Le Jeune, 1672). On the significance of Rohault's text, see Paul Mouy, *Le developpement de la physique cartesienne, 1646–1712* (New York: Arno Press 1981).

52. Jacques Rohault, *Entretiens sur la philosophie* (Paris, 1681). On Rohault's dialogue, see Van Damme, *Descartes,* 37–47.

53. Fontenelle, *Conversations,* 4

54. Ibid.

55. Ibid., 4–5.

56. Ibid., 5.

57. Ibid., 10.

58. "Fragmens d'un Traité de la raison humain," *OF,* 7:479.

59. Fontenelle, *Conversations,* 15.

60. Note that I disagree with the prevailing view that situates Fontenelle as an heir to Descartes and Cartesianism; see, for example, Louis Carré, *La philosophie de Fontenelle, ou La sourire de la raison* (Paris: Felix Alcan, 1932). Leonard Marsak, *Bernard de Fontenelle: The Idea of Science in the French Enlightenment* (Philadelphia: American Philosophical Society, 1959) offers a more nuanced account but still concludes by linking Fontenelle to French Cartesianism. Alain Niderst's view of Fontenelle, which stresses his attachments to Gassendi more than Descartes, is closer to my own. In addition to his biography of Fontenelle cited in note 19, see *Fontenelle à la recherché de lui-meme (1657–1702)* (Paris: A. G. Nizet, 1972). See also J. B. Shank, "On the Alleged Cartesianism of Fontenelle," *Archives internationales d'Histoire des sciences* 53 (2003): 139–56.

61. "Fragmens," *OF,* 7:477

62. Fontenelle, *Conversations,* 34.

63. Ibid., 65.

64. Ibid., 73.

65. Ibid., 10–11.

66. Ibid., 17.

67. Ibid., 21–22.

FRANCO ARATO

# Minerva and Venus

## *Algarotti's* Newton's Philosophy for the Ladies

Is it possible to explain natural sciences to everybody? To eighteenth-century women too? Voltaire was sceptical about it, making an exception for Madame Du Châtelet, his Muse and teacher who had studied Newton extensively (and for whom he wrote in his Newtonian treatise: "true philosophy is for every class and every sex"). But Voltaire lost his temper when a bookseller in Amsterdam published without permission his *Éléments de la philosophie de Newton,* adding the catching phrase *"à la portée de tout le monde"*: "You must be a charlatan," he wrote to Berger in 1738, "to add: *within everyone's reach,* and stupid *[un imbécile]* to think that Newton's philosophy could be for everybody."[1] Should Francesco Algarotti then, an intimate friend of Voltaire and of Madame Du Châtelet, also deserve the epithet of *imbécile* for writing *Newtonianismo per le dame (Newton's Philosophy for the Ladies),* a dialogue between a bright, learned man, with a sort of fanaticism for England, and a charming, young marquise? "Ladies," we have to remember, was the eighteenth-century equivalent of "everybody," for women at that time were usually banned from advanced studies.

Certainly the young Algarotti (born in Venice in 1712) was neither an illiterate nor a charlatan: he had been studying Newton for many years with Voltaire (at Cirey in the fall of 1735), who presented him to a friend as "a young man who knows the languages and habits of every country, who writes rhymes as Ariosto did, and understands Locke and Newton."[2] But a competition was bound to arise between the great poet and playwright and the novice writer, even if Voltaire generously welcomed the book (published in Milan in 1737) with flattering, perhaps *too* flattering, verses: "They praised your houses built on water, /

and your works are firmer than stone: / Venice and your book seem Gods' presents / but the last one will be more appreciated."[3]

Basically here were different conceptions about the nature of philosophical and scientific language. Voltaire employed in his *Éléments* (published in 1738, a few months after the *Newtonianismo*) an unadorned style that reads like a compendium for students. (He called it ironically *"mon catéchisme,"* implying perhaps that people always need a belief.) Algarotti imitated, or rather copied, as his Italian enemies said, the seductive model of Bernard le Bovier de Fontenelle's *Entretiens sur la pluralité des mondes habités (Conversations on the Plurality of Worlds),* first published in 1686 with many subsequent editions. He dedicated the book to the aging Fontenelle (*académicien* and, above all, *Cartésien*), invoking for his country a radical reform of culture and language and presenting to them the model of French *sociabilité:* "The Century of Things may come at last, for us too: and learning may not make harsher our minds or make us quibble about an old-fashioned sentence but may refine us and, if it is possible, embellish our society."[4]

Science in dialogue form had a long, authoritative tradition, as Galileo's great example makes clear. Truth, given in the Socratic way, arises from contradiction, and even in mathematics questions and answers are welcome, as in the last masterpiece by Galileo on mechanics and dynamics, *Discorsi e dimostrazioni matematiche intorno a due nuove scienze* (Discourses and Mathematical Demonstrations Concerning Two New Sciences) (1638). But in the Rococo era the arrival of women in the world of salons, and even of *académies,* transformed the sober dialogue into a social event, the formal ceremony of question and answer into a strange rite of *chuchotis et compliments,* of whispers and compliments. If women were absent in Galileo's dialogues (as in the male universe of Plato where only a fictitious Diotima appeared), the *esprit* of women was often required in eighteenth-century works of natural philosophy. The Italian literary academy of Arcadia, founded in 1690 almost as a posthumous tribute to Queen Christina of Sweden (who died in Rome as a Catholic), promoted women's learning. In the history of the academy (*L'Arcadia* [1708]), Giovanni Mario Crescimbeni described a party of so-called *pastorelle* (aristocratic "shepherdesses," as the curious rules and nicknames of Arcadia required) intensely interested in experiments with Boyle's air pump. Some years later a disciple of Algarotti, the Venetian Eusebio Sguario, had to introduce women into his very technical treatise *Dell'elettricismo* (On Electricism) (1746) in order to make it more interesting to readers. For these Italian writers, "science" and "philosophy," as we understand the terms in contemporary language, must be explained to the general literate public, which was sometimes shocked by new ideas about Nature and doubtful of the religious orthodoxy of the philosophers.

Among the early poems written by Algarotti in the spirit of the Arcadia we find a Newtonian canzone (1732) for the young Laura Bassi, one of the first women to receive a doctorate in Bologna. Recent studies demonstrate that though Bassi later came to be identified as a Newtonian, her doctoral dissertation on optics was in fact still strongly influenced by Descartes' theory of tourbillons. Therefore, when Algarotti wrote to Bassi (imitating Greek syntax) "of the golden, / seven times simple light / the variously burning, mixed and yet pure colors *[de l'aurata / luce settemplice / i vario-ardenti e misti almi color],"* he probably anticipated the advent of Newtonianism, not yet well established in Bassi's alma mater.[5] Bassi, like her contemporary Madame Du Châtelet, had access to mathematical studies and corresponded with the most important scientists and philosophes of Europe (including Voltaire). Unlike Du Châtelet, Bassi was given institutional recognition and became a professor at the Istituto of Bologna. In Milan, some years later, a true *enfante prodige,* Maria Gaetana Agnesi, began to amaze her masters by her skillfulness in mathematics. In 1748 Agnesi, only twenty-one, published her *Instituzioni analitiche ad uso della gioventù italiana* (Analytical Institutes for the Use of Italian Young People), dedicated to her patron Empress Maria Theresa and a major contribution to the diffusion of infinitesimal calculus in Italy. Agnesi justified the use of the Italian language instead of Latin; she mentioned the example of "many famous mathematicians living beyond the Alps [i.e., French]," and reminded her readers that her aim was clarity rather than "purity of language."[6] In Agnesi's treatise we do not find any of Algarotti's tricks and *tournures:* she spoke with a refreshing voice to young people fond of study (*"alcuno de' miei minori fratelli* [some of my younger brothers]") but employed appropriate mathematical language. Thus, she proved that a young woman could write soberly as her male colleagues did: to use Voltaire's terms, true learning is for every class and every sex.

We know, in contrast, that Algarotti's touch made Newton's physics into a play of society where the laws of attraction were explained with the language of *billets doux:* "after eight days' absence Love has decreased sixty-four times from the first day" (Voltaire doubted that this *calembour* could please the *"esprits bien faits"*).[7] And Voltaire, who was foremost a writer and not a scientist, initially had been shocked by the Newtonian *vague* and wrote to his friend Cideville: "Poetry is no longer fashionable in Paris. Everybody wants to be a mathematician and a scientist; Reason is everywhere, imagination and the Graces are banned." Later observing Algarotti's success, he complained to Thieriot: "I think that there are more truths in ten pages of my treatise than in his entire book: and this will perhaps overwhelm my book and make his success. He picked the flowers and left me the thorns."[8]

Algarotti was interested in showing how deep the revolution bound by the Newtonian system was: he looked to society as well as to laboratories, to amateurs as well as to professionals. If he exaggerated in his imitation of the style of the venerable Fontenelle (who, after all, in his *Entretiens* cared more about cosmological imagination than about reality), he was also an experimental natural philosopher, who as a young student in June 1728 had successfully repeated at the Istituto delle Scienze of Bologna the Newtonian *experimentum crucis* on the immutability of light. Many other Italian scientists and dilettanti before him had failed in this proof (including the Bolognese Francesco Zanotti, one of Algarotti's masters), because they used imperfect prisms.[9]

Algarotti was not the first Italian to perform the main experiments illustrated in Newton's *Opticks*. He was preceded by the erudite Francesco Bianchini in Rome and by the polymath Celestino Galiani in Naples (uncle of the more famous abbé Ferdinando),[10] but he was the first one to make public his demonstrations. His literary background was strongly French (influenced not only by Fontenelle and Voltaire but also by Montesquieu: he imitated the erotic *Temple de Gnide* in his *Congresso di Citera* [1745]). His philosophy was English: Locke; Bolingbroke; and the classical scholars William Temple, with his preference for Epicurus, and Conyers Middleton, the biographer of Cicero. Locke gave Algarotti the means to understanding not only the laws of knowledge but also the concept of free will (the "uneasiness": a little word that had even before Algarotti tormented the Christian philosopher Ludovico Antonio Muratori, the founding father of Italian historical erudition). Locke, as we know, was belatedly prohibited by the Catholic Church in 1734—forty years after the publication of the *Essay Concerning Human Understanding*!

Algarotti, in the different editions of his book, never included the metaphysics inspired in England (and later in Italy) by the Newtonian cosmology (the so-called Boyle lectures). Such religious interpretation, that we might include within the term "natural philosophy" (and that finds some justification in the famous *Principia*'s *"Scholium generale"*), was still attractive, however, for Voltaire, who in the second edition of his book on Newton began, as in some seventeenth-century treatises, with a chapter about God: "Many people will be perhaps surprised that among the proofs of God's existence, the strongest in Newton's opinion was the argument of the final cause."[11] Even so, Algarotti found the way to express some doubts about the metaphysical implications of Newton's universe: "Newtonian attraction," he wrote, "is nowadays put alongside with efficacious and effectual Grace: I think these entities would be very surprised to have such an encounter."[12] Algarotti in 1737 was so eager to receive the applause of his readers[13] that he underrated the threat of ecclesiastical condemnation by refusing to give the usual (and hypocritical) anti-Copernican declaration. The prohibition of Algarotti's *Newtonianismo* arrived soon after publication (April 1739), despite the fact that the Congregazione dell'Indice was headed by Cardinal Giovanni Antonio Davia, a great friend of the group at the Istituto of Bologna (and a dilettante scientist himself). Perhaps as old bloodhounds the majority of the members of the Congregation smelt in the younger writer a free thinker who (probably) was associated with the Masonic lodge in Florence. In Florence, Algarotti had known the vice-president of the British Royal Society in Newton's time, Martin

Folkes. During his own stay in England in 1736, Algarotti had become the friend of John Theophilous Desaguliers, the Grand Master of the Masonic lodge in London.[14]

In fact, Italian scientists had to wait for the "liberal" Pope from Bologna, Prospero Lambertini (Benedict XIV), who in the 1740s encouraged informed opinion, to separate the old Copernican (and Galileian) question from the study of Newtonian optics and cosmology. The great edition of *Principia* published by the Minim friars Le Seur and Jacquier (Geneva, 1739–1742) was prepared in Rome, in the convent of Trinità dei Monti.[15] Some years later the Jesuit Ruggero Boscovich became one of the most influential Newtonian scholars in Europe and, as a man of literary good taste, imitated the Algarottian manner somewhat in his *Dialogi sull'aurora boreale* (Dialogues on Aurora Borealis) (1748) but, following the Jesuit tradition, without admitting any woman into the imaginary conversation.

In the meantime Algarotti worked hard to revise his book.[16] The last version (1764), with the less brilliant title of *Dialoghi sopra l'ottica neutoniana* (Dialogues on Newton's Optics), is almost completely free of polemical stings; the author also tried to abandon the imitation of French style, sometimes adopting the cold "righteousness" of linguistic *purismo,* reminiscent of Boccaccio's *Decamerone*. The condemnation of the Church had some effect, even in the history of the Italian language.

We can go further in comparing Voltaire's *Éléments* and Algarotti's *Newtonianismo*. Voltaire's book is a complete Newtonian handbook, with questions and answers. Algarotti, in contrast, chose to escort his marquise protagonist of the dialogues (Madame Du Châtelet briefly thought herself to have been the model for this capricious character) through all the absurdities of Descartes' and Malebranche's optics, in order to provoke a spontaneous rejection of these models in favor of Newtonianism. This seems to be a sort of *itinerarium mentis in veritatem* (to borrow a famous expression from the medieval friar Bonaventure), conducted with the constant help of prisms: not a dogmatic lesson, but an essay in the inductive process. Language had to be adapted to impart the few pieces of information for the common reader and (what constitutes the charming yet irritating aspect of the book) for the gallantry of the salons and boudoirs: "*instruire l'esprit en parlant toujours à l'imagination* [to instruct the mind, while always speaking to the imagination]," as Algarotti wrote in his dedication to Frederick II King of Prussia.[17]

Nevertheless, we find striking similarities between Algarotti's and Voltaire's books: in the introduction a short history of physics, and optics in particular, starting with the remote medieval origins (and here the Italian writer had an easy time defending two of the so-called *martiri della Ragione* [martyrs of Reason], Roger Bacon and Galileo).[18] In general, both Algarotti and Voltaire were sceptical of humans' ability to understand the ultimate laws of the universe. They accepted universal attraction as a fact, not as an occult quality: here they seem to share with modern scientists the tendency to avoid metaphysics and to define the limits of demonstrable knowledge. Let us compare two very similar pages from Algarotti and then from Voltaire:

> Attraction, far from being an occult quality, is a very evident property of matter: on it diffraction and refraction evidently depend, and many other phenomena; attraction is by no means a name without content, invented just to explain some of Nature's manifestations, but a general principle acting on the grain of sand as well as on the largest of our planets. . . . We have to remember that Newton when he says that light, passing near to a body, is attracted by this same body, does not want to give a complete explanation of the phenomenon, but only to point out a property of mass on which diffraction depends, the ultimate cause remaining unknown.[19]

> They often say that attraction is an occult quality. If they mean by this word a law that we cannot justify, the whole universe is in this condition. . . . If they mean a locution of the old Scholastica, a word without content, let us consider that this so-called chimaera is one of the most sublime and precise mathematical demonstrations that Newton disclosed to mankind. We saw that rays reflected by a mirror cannot come back to us from the mirror's surface. We tested that rays transmitted through a prism following a definite angle come back instead of passing in the air, and we saw that if there is void behind this prism, rays that were transmitted before, turn back to us. Without any doubt there is no known impulsion here.[20]

As the last editors of Voltaire's *Éléments* have pointed out, Algarotti probably suggested to Voltaire the way to use "optics to present universal attraction,"[21] an aspect rather unusual in the popularization of Newtonianism. (Newton himself had confined the important subject to the last "Queries" of the third book of his *Opticks*.) Algarotti showed then a childlike enthusiasm for the futuristic project of the Jesuit Louis Bertrand Castel who, speculating on the parallelism between the diatonic and the twelve-note chromatic scale, announced the construction of a *clavecin oculaire*. "To the touch of keys," wrote Algarotti, "instead of listening to a sound, you will see colors and half-shades, that will have among them the same harmony of sounds."[22] Voltaire too, though with less enthusiasm, followed Algarotti's example and paid attention to the "virtual" invention of the brilliant Father Castel.[23]

Anti-Cartesian polemics are at the centre of Algarotti's interest, while Voltaire preferred to edit out gradually his harsh words against the French philosopher. In fact, the Venetian writer had to struggle at home against "*l'idra a cento teste* [the hundred-headed hydra]" of Cartesianism, one of these heads being the dilettante Giovanni Rizzetti (from Bergamo), who wrote against Newton for the *Acta Eruditorum* of Leipzig. Rizzetti's dubious theories, based on the subjectivity of perception and partially inspired by an old book by the French mathematician Edme Mariotte, are nowadays almost forgotten by scientists. They would be brought to life again in Goethe's *Farbenlehre* (Theory of Colors). (Goethe, as is well known, considered Newton's mechanicism cold and lacking in poetry.)

Algarotti's *Newtonianismo* had a great success, as testified to by numberless translations and imitations. The French translator, the diplomat and novelist Louis Duperron De Castera (*Le newtonianisme pour les dames* [Paris, 1738]), was a very unhappy choice, as his knowledge was limited both in physics and in the Italian language. "He translated my book," wrote Algarotti to Voltaire when invoking his help, "in the taste of this English writer who translated *esprit bas* as 'a wit of the stocking'" (*'bas'* in French means 'low,' as well as 'stocking').[24] A Russian version by the influential poet and scholar Antiokh Cantemir (Algarotti visited Russia in 1739 and wrote a very interesting travel book) was announced, but never appeared in print, though Cantemir, who had already translated Fontenelle into Russian, as far as we know completed the work, but the manuscript has been lost.[25] The English version by Elizabeth Carter (a friend of Samuel Johnson) was a far better achievement, but it did not prevent a merciless judgement by Horace Walpole: "Algarotti would dishearten anybody from attempting to meddle with the system of the universe a second time in a genteel dialogue."[26]

Newton's optics and physics at last triumphed in Italy in the second half of the eighteenth century. Algarotti, who had been in a certain sense a pioneer, at least in the new frontier of sociability, was frequently considered after his death only a superficial latecomer. And speaking about love in the language of physics became a habit to ridicule. The great Milanese poet Giuseppe Parini, an admirer of Pope's *Rape of the Lock,* in his satirical poem *Il giorno* (published in 1764, the year of Algarotti's death) described and mocked a situation that we can easily compare with some of the images from Algarotti's dialogues:

> Let Madame with a desiring look
> and an approving nod of her charming head
> listen from your lips
> the *calculus* and the *mass,*
> and you please make resound again
> the *inverted reason* in your lovely mouth.[27]

But among the same aristocrats in Milan, described so sentimentally by Parini, science was a symbol not only of fashion but also of intellectual freedom. In 1764 Pietro Verri, the leading figure of the Italian Enlightenment, in the avant-garde and learned newspaper *Il Caffè* spoke of Algarotti and of his friend, the Tuscan medical doctor Antonio Cocchi (another Freemason and anglophile), as writers who "in different ways but with the same philosophical spirit have enriched our language and have left to us books full of noble and great ideas, adorned with a style that makes them even more charming."[28] For Verri, Algarotti was legitimately one of Galileo's heirs, since, as the great master had done in his *Dialogo sopra i due massimi sistemi del mondo (Dialogue Concerning the Two Chief World Systems),* he had tried to persuade his readers to accept the reasoning of science in place of superstition and philosophical dogma. Obviously Algarotti's style and genius, as noted previously, were very different from Galileo's. In the age of Enlightenment, access to learning meant a much wider and, at the same time, less demanding audience. As a result, in Italy as elsewhere every popularization of science became, as Algarotti himself well knew, a risky task.

If we take a look today at the everyday style of scientific popularization, so necessary in our age of dogged specialization, we have some difficulty in accepting Algarotti's prose as modern, at least in comparison with Voltaire's classical style. *Clarté* (clarity) seems to be the imperative of the scientific writer. But we find some exceptions: for example in the mesmeric, if not tantalizing, work of a scholar from the United States on artificial intelligence, Douglas R. Hofstadter. In his *Gödel, Escher, Bach* (1979), he combined algorithms with witty dialogues written in imitation of Lewis Carroll's tales; in a more recent book the same author tried to explain the philosophical and philological riddle of translation using scientific *metaphorae* (*Le Ton beau de Marot* [1997]): once again mathematics goes along with literature, in the best Anglo-Saxon tradition. We are perhaps not so far from Algarotti's eighteenth-century inquiring marquise, who claimed the value of the combination of imagination and learning in the same way.

## Notes

1. Voltaire to Berger, 14 May 1738, D1502. Quotations are always from Theodore Besterman's definitive edition of Voltaire's *Correspondence and Related Documents* (Oxford-Genève-Toronto: Voltaire Foundation, 1968–1977); translations are mine. Voltaire's path to Newton is excellently explained in Voltaire, *Éléments de la philosophie de Newton,* ed. R. L. Walter and W. H. Barber, vol. 15 of *The Complete Works of Voltaire* (Oxford: Voltaire Foundation, 1992), 29–58; the third edition of 1741 is here reprinted with a choice of author's variants. A sentence at the beginning of the Dedication of the *Éléments,* "Ce n'est point ici une marquise, ni une philosophie imaginaire . . . [I don't speak here about a marquise, or an imaginary philosophy . . .]," should be read as an implicit criticism of Algarotti.

2. Voltaire to Thieriot, 3 November 1735, D940. For the personal relationship between Algarotti and Voltaire, see H. T. Mason, "Algarotti and Voltaire," *Rivista di letterature moderne e comparate* 33 (1980), 187–200.

3. These verses appeared in the second edition of Algarotti's *Newtonianismo.* See Francesco Algarotti, *Il newtonianismo per le dame, ovvero dialoghi sopra la luce, i colori, e l'attrazione* (Napoli [*recte:* Venice, Pasquali], 1739).

4. Ibid. In the frontispiece, the reader could find, along with an elegant engraving from the Venetian painter Giambattista Piazzetta, a quotation from Vergil's tenth eclogue ("quae legat ipsa Lycoris [Let Lycoris too read this]"). Voltaire, who disliked the Dedication to Bernard de Fontenelle, used his bitter irony in commenting on the quo-

tation in a letter to Thieriot (18 May 1738, D1505): "*Nota bene* que *quae legat ipsa Licoris* [*sic*] est très joli, mais ce n'est pas *pauca meo Gallo* [Cornelius Gallus mentioned in the Vergilian eclogue]; c'est *plurima Bernardo* [Bernard de Fontenelle]." Some years later, in 1750, Algarotti chose to write an opportune new Dedication to Frederick the Great, King of Prussia, his patron since 1740. See also Philippe Hamau, "Algarotti vulgarisateur," *Cirey dans la vie intellectuelle SVEC* 11 (2001): 73–89; and Massimo Mazzotti, "Newton for Ladies: Gentility, Gender and Radical Culture," *British Journal for the History of Science* 37 (2004): 119–46.

5. On Bassi and Algarotti, see Marta Cavazza, *Settecento inquieto. Alle origini dell'Istituto delle Scienze di Bologna* (Bologna: Il Mulino, 1990), 237–56. It is interesting to point out that in 1723 the great biologist Antonio Vallisneri wrote a short memoir for the "Accademia de' Ricovrati" in Padua entitled *Se le donne si debbano ammettere allo studio delle scienze e delle arti nobili* (If Women Ought to Be Allowed to Study Sciences and Noble Arts), where he speculated, in a rather discriminatory way, that scientific researches should be carried out only by women "of noble and illustrious blood, whose mind shines and glitters in an unusual degree, far beyond the range of the common herd." *Opere fisico-mediche stampate e manoscritte del Kavalier Antonio Vallisneri raccolte da Antonio suo figliuolo* (Venezia: Coleti, 1733), vol. 3, 603. See also Paula Findlen, "Becoming a Scientist: Gender and Knowledge in Eighteenth-Century Italy," *Science in Context* 16 (2003): 59–87.

6. Maria Gaetana Agnesi, *Instituzioni analitiche ad uso della gioventù italiana* (Milano: Nella Regia-Ducal Corte, 1748), vol. 1, "To the Reader." On Agnesi, see M. L. Altieri Biagi and B. Basile, eds., *Scienziati del Settecento* (Milano-Napoli: Ricciardi, 1983), 755–78. Agnesi's textbook was translated into English by J. Colson and into French by P. Th. Anthelmy.

7. Voltaire to Maupertuis, 22 May 1738, D1508. Algarotti's quotation from *Newtonianismo,* 1st ed. (1737), 250 ("Dopo otto giorni di assenza l'Amore è divenuto sessantaquattro volte minore di quello che fosse nel primo giorno").

8. Voltaire to Cideville, 16 April 1735, D863; Voltaire to Thieriot, 18 May 1738, D1505. The Newtonian fashion in poetry and arts was documented by the innovative book of Marjorie H. Nicolson, *Newton Demands the Muse: Newton's Opticks and the Eighteenth Century Poets* (1946; repr., Westport, CT: Greenwood Press, 1979).

9. See *De Bononiensi Scientiarum et Artium Instituto atque Academia Commentarii* (Bologna: Dalla Volpe, 1731), vol. 1, 200–201; Franco Arato, *Il secolo delle cose. Scienza e storia in Francesco Algarotti* (Genova: Marietti, 1991), chap. 1. For Algarotti's biography, see Ida F. Treat, *Un cosmopolite italien du XVIIIe siècle: Francesco Algarotti* (Trévoux: Jeannin, 1913); and Ettore Bonora, "Francesco Algarotti," in *Dizionario biografico degli italiani* (Roma: Istituto dell'Enciclopedia italiana, 1960), 2:356–60. A modern choice of texts is available in Algarotti-Bettinelli, *Opere,* ed. E. Bonora (Milano-Napoli: Ricciardi, 1969). For the history of Bologna Istituto delle Scienze, see W. Tega and A. Angelini, eds., *Anatomie accademiche* (Bologna: Il Mulino, 1986–1993), 3:313–409. There is also a very useful register of the physical experiments performed at the Institute in the eighteenth century.

10. For the history of early Italian Newtonianism, see Salvatore Rotta, "Francesco Bianchini," in *Dizionario biografico degli italiani* (Roma: Istituto dell'Enciclopedia italiana, 1968), 10:187–94; Vincenzo Ferrone, *Scienza natura religione. Mondo newtoniano e cultura italiana nel primo Settecento* (Napoli: Jovene, 1982); Paolo Casini, *Newton e la coscienza europea* (Bologna: Il Mulino, 1983), 173–227; Paolo Casini, "The Reception of Newton's *Opticks* in Italy," in *Renaissance and Revolution: Humanists, Scholars, Craftsmen, and Natural Philosophers in Early Modern Europe,* ed. J. V. Field and Frank A. J. L. James (Cambridge: Cambridge University Press, 1993): 215–27.

11. Voltaire, *Éleménts de la philosophie,* 197. "Plusieurs personnes s'étonneront ici peut-être, que de toutes les preuves de l'existence d'un Dieu, celle des causes finales fût la plus forte aux yeux de Newton."

12. Letter to the Marquis N. N., 4 April 1763, Francesco Algarotti, *Opere* (Venezia: Palese, 1791–1794), vol. 10, 72.

13. In an amusing letter to his brother Bonomo (4 December 1737), Algarotti compared the efforts of modern writers to please the public to the tender struggles in a love affair: "Authors are like lovers. Their capricious and difficult Lady is the severe public. They pay attention to everything and they think that the lightest gesture or glance can change their destiny," quoted from the manuscript in Arato, *Il secolo delle cose,* 30.

14. See Mauro De Zan, "*La messa all'Indice del* Newtonianismo per le dame," in *Scienza e letteratura nella cultura italiana del Settecento,* ed. R. Cremante and W. Tega (Bologna: Il Mulino, 1984), 133–47.

15. See Casini, "Reception," 223.

16. For a bibliographical survey of eighteenth-century editions and translations of Algarotti's book, see Arato, *Il secolo delle cose,* 133–55.

17. Francesco Algarotti, *Dialoghi sopra la luce, i colori, e l'attrazione* (Berlin: Michaelis, 1750); it is an extensively revised edition of *Newtonianismo,* "Au Roi." In the same French dedicatory letter we find some interesting observations on the origins of the modern Italian language: "Our language is, I daresay, neither alive nor dead. We have some authors of the remote past we consider our classics: but their books are full of affectation and of out-moded words."

18. See Algarotti, *Newtonianismo,* 1st ed. (1737), 15, 117–18.

19. Ibid., 230.

20. Voltaire, *Éléments de la philosophie,* 345–46.

21. Walter and Barber, introduction to *Éléments de la philosophie,* 45; Walter and Barber mention here some of the rare expressions of Du Châtelet in favor of *Newtonianismo.*

22. Algarotti, *Newtonianismo,* 1st ed. (1737), 138. This multimedial eighteenth-century dream (or *Gesamtkunstwerk* dream) did not encounter any success, notwithstanding the realization of some prototypes. See bibliography in A. Cohen, "L. B. Castel," in *The New Grove Dictionary of Music and Musicians,* 3:865.

23. See Voltaire, *Éléments de la philosophie,* 393–94.

24. See Algarotti's letter to Voltaire, 27 September 1738, in *Giornale storico della letteratura italiana* 164 (1987), 565–67. I mention only two examples: where Algarotti wrote "stelle inerranti e fisse [not wandering and fixed stars]," Castera translated "étoiles tant fixes qu'errantes"; where the Italian, speaking about Castel's harpsichord, made a playful allusion to the future, "musica delle salse [the music of sauces]," Castera wrote "peut-être qu'un jour le clavecin nous donnera de quoi dîner [someday perhaps the harpsicord will give us the way to make a living]." Voltaire and Du Châtelet declined Algarotti's request to collaborate in a new translation, though Voltaire did encourage a new young protegée to undertake the task. A good French translation appeared after Algarotti's death in the edition of the *Oeuvres* sponsored by Frederick II (Berlin, 1772).

25. See Valentin Boss, *Newton and Russia: The Early Influence, 1698–1796* (Cambridge, MA: Harvard University Press, 1972), chap. 13.

26. Carter's translation appeared with the title *Sir Isaac Newton's Philosophy Explained for the Use of the Ladies* (London: Cave, 1739). Walpole's quotation from Horace Walpole to William Mason, 1782, *Letters of Horace Walpole,* ed. Paget Jackson Toynbee (Oxford: Clarendon Press, 1910), 12:173.

27. Francesco Algarotti, vv. 983–88. For subsequent images of women, see Rebecca Messbarger, *The Century of Women: Representations of Women in Eighteenth-Century Italian Public Discourse* (Toronto: University of Toronto Press, 2002).

28. *Il Caffè* (1764), 1, 19.

## **SECTION III:**

## WOMEN, MEN, AND THE NEW SCIENTIFIC ESTABLISHMENT

LYNETTE HUNTER

# Women and Science in the Sixteenth and Seventeenth Centuries

## *Different Social Practices, Different Textualities, and Different Kinds of Science*

The impetus for this essay came from recognizing that men and women practised science in the same places and with roughly the same equipment up until the middle of the seventeenth century. However, they practised science for different reasons, leading them to communicate in different ways, and these different rhetorics have had a long-term impact on access to scientific power and to the legitimation of particular methodologies and of various kinds of scientific knowledge. The focus of this essay is on the modes of communication used by women for sharing and preserving knowledge, and the impact that these modes had on the kind of knowing and the kind of science in which they engaged. Their practices ran parallel to those of modern science and suggest ways of engaging with the natural world that were not recognised by the scientific community at the time, and which we still have difficulty valuing.

One significant, but taken for granted, aspect is that women were not, with rare exceptions, involved in generating "natural philosophy" in the eras before the fifteenth century. Natural philosophy was primarily communicated through oral disputation with formal rhetorical rules, accompanied by some manuscript circulation and commentary. The rhetoric of women's communication about interaction with the physical world, which pertained directly to their responsibilities for domestic medicine and household technology, may have been oral, but there is no record of formal disputation or their education in it. When men of the seventeenth century brought together scientific debate with practical experiment, it is arguable that natural philosophy gave way to science and philosophy as distinct areas of knowledge. At this time, the growing number of educated women could have introduced a written rhetoric more appropriate to their own practice. However, they did not, and the trajectory of this essay attempts to understand why.

One primary question is: why with so many women having a considerable reputation for the practice of experiment, and with so many of them closely associated with the men who formed the Royal Society, were they then excluded from meetings and communications? The simple fact is that, for example, one of its predecessors, the Hartlib circle,[1] included many women who corresponded with others, both men and women, about medical remedies, agriculture, botany, chemistry, and pharmacy. It was undoubtedly the case that some women were accorded more respect than others. In the early years of the circle, Joan Barrington predominated, in the 1640s it was Dorothy Moore, and in the 1650s it was Katherine Boyle.[2] Their participation was marked by manuscript culture, and the cessation of Hartlib's copyhouse in 1660 may distort the actual events. Nevertheless, that so many men of the circle and their close associates went on into the Royal Society, which excluded these women, must have changed the way in which women in general were perceived and in which they perceived their own contributions.

From my own close research activities alone[3] we find that Katherine Boyle (married name Katherine Jones, Lady Ranelagh) probably knew most of the membership but was particularly close to her brother Robert Boyle, to Richard Jones, to John Beale, and to Thomas Willis. She and her brother lived in the same house in London in their latter years, a house which held his laboratory.[4] Her associate and friend Dorothy Moore (born Dorothy King, later to become Dorothy Dury)[5] corresponded with William Petty, John Pell, Henry Oldenburg, and John Clotworthy. Alethea Talbot (who later became the Countess of Arundell and Surrey) had worked with many, including Elias Ashmole, and was the mother of Henry Howard. Her sisters Mary and Elizabeth, with their friend Anne Clifford, had had close connections with Gilbert Talbot, John Aubrey, and William Cavendish. And of course there are many more connections,[6] including the mother/son link between Joan and Thomas Barrington[7] and the husband/wife relationship of Mary and John Evelyn.[8] Many of these women were not only in the Hartlib circle[9] but were also associated with the group at Wadham College, Oxford,[10] and peripherally, through Kenelm Digby, with Gresham College.[11]

The women who were being praised for their scientific judgement until the end of the 1650s by the very men who established the Society may have been excluded because of a need to define gender characteristics in a different way. After all, the seventeenth century in England saw the construction of the domestic housewife, that radically new imagined entity, the unpaid shadow necessary to a middle-class way of life. It would have been difficult to differentiate between the abilities of women and men on theological grounds, and in most cases the aristocratic titles of these women would have made it impossible on the grounds of status. I would like to suggest that the reason women were excluded from the Royal Society was largely because of their social practice of science, which was tied to a local community.

While most of the evidence for the argument that follows is drawn from the written and manuscript material left by gentlewomen and noblewomen, I would argue that their common basis in communal activity provides grounds for a reasonable hypothesis that other women in the community were practising science in the same way. Curiously, it may also be the case that the need to redefine gender derived from an urge to democratise. Because the new practices and the rhetoric used by men of the Royal Society legitimated their social practice as part of the new proto-liberal social contract, the product of the Civil War and the events surrounding the coronation of William and Mary in 1688, the social practice of science by women was placed outside England's new definition of the democratic citizen and the practices associated with it.

## Rhetorical Elements in the Practice of Science for the Men of the Royal Society

The practice of science for men in the Royal Society came to be defined by two primary rhetorical elements that distinguished it from the practice of science by men preceding the Royal Society. First, in the Society itself and in its published *Transactions,* scientific practice consisted of a public display of intervention into nature, by way of visual demonstration and written representation of that intervention or experiment, so that it could be repeated.[12] In this, Royal Society practice appears at first sight not much different from the social practice of scholastic and early Humanist *scientia naturae,* which worked from supposition, through demonstration, in the oral rhetoric of disputation. Yet despite the similarities supposition is not analogous to hypothesis. Neither is the Aristotelian demonstration derived from the *Posterior Analytics* analogous to visual demonstration.[13] Nor is the oral rhetoric of disputation analogous to written representation. However, the fundamental difference is that for a medieval world, interventions into nature were correspondent rather than actual. The medieval scientist was interpreting God's structure, which was unfathomable. Even Aquinas never went beyond the doctrine of suppositional necessity and probability. In contrast, the Royal Society scientists were dealing with the possibility of actual intervention, from which ultimately would be derived the certainty and universalism of modern science.[14] Where the two kinds of scientific practice were similar is in their social structure, especially in terms of the self-contained, end-directed rhetoric of club culture fostered by each.[15]

The second element defining the rhetoric of Royal Society practice came from that specific intervention in the actual. This was carried out initially in laboratories, but more often at home in kitchens and stillrooms—places quite private compared to the end result of public demonstration. In this, Royal Society practice was again similar to the earlier practices of artisans and tradespeople working on "secrets," which were also intensely private even if always without public demonstration, to avoid intellectual theft.[16] But the fundamental difference between the practice of these earlier experimenters and that of the early modern scientist, was the need to link private experiment with public discussion: to link the intervention in the actual with public display of that intervention in visual demonstration and written representation.

Why was there this importance to link the private practice of intervention with public display and discussion? Largely, I suggest, because the men involved saw a need, theological, political, and ethical, for greater access to the "secrets" of nature. But as this paper will go on to explore, women practising science had already developed a procedure for exchanging information about scientific knowledge, so why did the men not follow their specific mode of communication? Part of the answer is that men in the Royal Society were differently organised than women and needed different rhetorical structures for communication. There were more of them who perceived themselves to be part of a community, and that community was not local, but regional, national, and international. Also, the men often wished to exchange information across larger geographical space; hence, they needed to write rather than develop knowledge by speaking to each other or by tacit observation. They perceived themselves, consciously or not, as class and gender unified, part of a much larger movement toward the unification of a particular class and a specific gender that was becoming represented in politics. The structure of communication in their modern science displayed the structure of the universal man/private citizen dichotomy made necessary by the evolving liberal social contract: representation had to be stable and repeatable to make the representative democracy of post-Hobbesian politics work,[17] just as it has to be stable and repeatable to make the experiment replicable. In other words greater access to the secrets of nature, for the men of the Royal Society, was a democratic necessity. Early modern science responded to this growing need for access, to a larger public, and to making knowledge public. And it answered that necessity by developing strategies that became simultaneously appropriate to political representation in the proto-liberal social contract.[18]

## The Social Practice of Science by Women

So in what sense was the social practice of science different for women? To understand some of this detail I need to look back into the sixteenth century, over 150 years marked out by the dates 1534, 1617, and 1649. Until the early sixteenth century women had been involved in the commercial practice and use of science, although this became increasingly difficult with the controls levied by guilds over trade[19] and by the Church over medicine in the late fifteenth century.[20] One large body of practice was in nunneries, which frequently had to engage in a self-sufficient scientific technology. Nuns, along with many women, practised medicine on Biblical authority, working in hospitals and almshouses with the signal role of touching the bodies of their patients, because physicians did not. Even women of the gentry and nobility were associated with hospitals, although more usually as patrons.[21] But also, all women had to practise some form of science daily, from the preservation of food to the production of cleaning agents and the maintenance of healthcare for people and animals, and there were many other practices which would now be called chemical technology. Women of the gentry and nobility had to oversee these practices if not perform them themselves:[22] we find this attested to in account books[23] but also implied by the architectural arrangement of the late-medieval house in which mistresses, masters, and their household members lived closely alongside each other, around the great hall.[24] Most of this practice was habitual technical performance, but from manuscripts we learn that several women at least engaged with and responded to changes in the environment, for example, learning how to deal with smallpox coming into England in the late fifteenth century. Although women did work with the commercial secrets of the artisan world, they had a much larger commitment to communal information—this kind of science being one of the primary modes for women to participate in "service" or public action in society.

The year 1534 marked the beginning of the dissolution of the Roman Catholic Church in England. A number of historians have linked the growing power of the Royal College of Physicians (incorporated in 1518) and the attendant difficulties for the poor to receive or obtain medical attention to the rescinding of the Quacks Charter in 1542–44. At the same time there was a sharp increase in the number of vernacular books on medicine addressed to yeomen and to the gentry from the 1530s onward—many of these with phrases such as "for the common good" or with references to the "commonweal(th)" embedded in the titles or addresses to the reader.[25] As important, if not more so, was the effect of the dissolution on the hospitals and almshouses run by the Church all over England, which had been particularly key to rural areas.[26] From the little research there is in the field come examples such as St. Leonard's in York, a huge hospital which was simply closed; St. Giles' in Beverley, to which the Earl of Rutland acquired patronage before abolishing;[27] and St. John's in Exeter, which was given in 1540 to Thomas Carew, who promptly turned it into a private dwelling.[28] Presumably because most of the evidence was burned when substantial buildings like Fountains Abbey were gutted, there is little paper evidence of the role of the monasteries and nunneries in interaction with local communities. However, I would hypothesise partly from the extensive trade that we know went on between the religious and lay communities, and partly from the tensions that resulted from the dissolution, that the religious houses had a substantial social role in their region, which probably included medical care and the provision of items of chemical technology either in practice or through patronage.

The dissolution also meant the redistribution of grounds and buildings into the hands of the new gentry and the new nobility: this in itself being one of the reasons for the plethora of books which acted as guides to behaviour when one became responsible to, or for, a community of people dependent on the lands. The main result of redistribution was the formation of the English country house phenomenon: the gentry and nobility decamped to the country house from the London court at various times during the year. The English country house operated a green, self-sufficient economy, often directly maintaining over one hundred to three hundred people and indirectly many more. They were large-scale businesses for the period. I would again hypothesise that some of these estates may have replaced the functions of the monasteries and nunneries in terms of healthcare and the provision of other household and agricultural products. While that is an hypothesis, we do know that from 1530 to 1580–90 a generation of gentlewomen and noblewomen became increasingly recognised for their skill in preparation of many household and medicinal chemicals and pharmaceuticals: Honor Lady Lisle;[29] the Cooke sisters, daughters of the tutor to Edward VI and Elizabeth, one of whom, Mildred, married Thomas Cecil, becoming the Countess of Essex, whose household was described as a "domestic university";[30] and Ann Dacre, the wife of Thomas Howard, Duke of Norfolk and Surrey, whose son Thomas married Alethea Talbot and whose grandson Henry became a member of the Royal Society.[31]

The work of these women is relatively well known; what I would like to stress here is the communal element of their work, the way they integrate it into service for their household and local area. And there is evidence of a class continuum of this kind of science practice, including not only the nobility but also the gentry and the emerging middle-class housewife. Thomas Tusser's *100 Pointes of Good Huswiferie* (1557), dedicated to Lady Elizabeth Paget, stresses the need for any housewife to know how to carry out basic preparations, especially of herbs,[32] and to do surgery.[33]

By the end of the sixteenth century, there were many women practising science and medicine, but the practice had diversified. No longer was it only from necessity or from service, but it had developed into a leisure activity. The women of the gentry and aristocracy practising between 1590 and 1649 include the relatively well-researched Ann Clifford, Margaret Hoby, Grace Mildmay, Alethea Howard and Elizabeth Grey (the Talbot sisters, whose third sister, Mary, was married to William Cavendish), Margaret Duchess of Cumberland, Joan Barrington, Brilliana Harley, and Queen Henrietta Maria, as well as many others.[34] Some practised only for what was needed for their families; others like Grace Mildmay produced on a commercial scale. They may have practised, as Hoby and possibly Mildmay and Barrington did, not only as a leisure activity but as religious and communal service.

Or they may have taken up practising as one of the indications of the status of noblewoman, as the distinctly short-lived nobility of the Talbot sisters' family may have decreed. Science, medicine, and chemical technology, including the use of new foodstuffs such as sugar—a notoriously difficult substance with which to experiment—seem to have become by the end of the sixteenth century markers of status. John Partridge's *The Widdowes Treasure* (1585) addresses the reader, saying that this kind of information is necessary to a particular kind of woman. They need to know how to perform these tasks in order to behave appropriately. In other words, the practices are a signal of conduct. There is evidence for a number of related books in the period having been written by women but midwived by men,[35] and books in these fields are clearly one of the most important genres for printing and publishing directed at women readers.

Increasingly, the idea of service as a responsibility of the aristocratic lady became redundant as the nobility crystallised in the reign of James I. Aristocratic households began to design and live in buildings that separated them physically from their retainers, who became "servants" rather than household members. Science began to be practised not only in the kitchen but also in the laboratory. For example, Mary Sidney (Mary Herbert) had her own laboratory, which was used by her brother Phillip and by William Cavendish, but there is little evidence that she was tending to a larger community. The development may partly be due to a shift to an urban environment in which these women did not have a distinct community.[36] It also may be partly a response by an increasingly leisured class to the question of what to do with their time. There may be an analogy between the circles of reading that developed among privileged women[37] and the groups of women experimenting (whose works are attested to by the attributions of recipes in manuscript writings).[38]

It is also quite possible that the practice of science was a permissible mode of intellectual interaction and of interaction with men since nearly all these women worked with men. Ann Dacre and Mrs. Dyce worked with Dr. Martin of Kornbeck (doctor to Henry VIII); Mary Sidney with her apothecary, Adrian Gilbert, and with Thomas Mouffet;[39] Elizabeth Grey with several people (according to Aubrey); Alethea Howard with individuals both at home and abroad as she found people with similar interests; Mildmay with her local doctor and apothecary; Barrington with Hartlib; and Henrietta Maria with Kenelm Digby and Thomas Mayerne. If one looks at Gerard's *Herball* or at any number of extant manuscripts, men and women working together seems to have been common practice among the gentry and yeomanry as well.[40] But the signal characteristic of these partnerships is the relative lack of aristocratic men.[41]

In 1617 a significant event took place. The Royal College of Physicians gained control of the *Pharmacopoeia Londoniensis,* the publication of the recipes used by the newly incorporated Society of Apothecaries. The control of the Apothecaries by the Physicians had begun in a slightly earlier period,[42] but 1616–17 was the first time that one finds warnings to women not to overstep their mark in books related to these areas,[43] and from 1617 to 1653 only one new book for women was published that was concerned with household science.[44] Furthermore, whereas in the earlier period we can often plot an interest in science from parent, especially mother, to daughter, we do not hear about the practice of the daughters of any of these women. It is probably significant that the women closest to the future members of the Royal Society in the 1640s and 1650s, Katherine Boyle and Dorothy Moore, were both from Ireland. Another hypothesis I would offer is that women in Ireland continued to be educated as their mothers had been a generation earlier, even though the fashion may have died out in England. Not only Katherine but also her sisters Mary and Lettice wrote and practised in the various areas of science, medicine, and household technology.[45] Dorothy Moore's sister Margaret Lowther was considered immensely learned, and her correspondents Margaret Clotworthy, Elizabeth Carey, and Ann Stanhope seemed to have had similar intellectual interests.

Not until after 1649 does this unofficial but effective prohibition on books for and/or by women cease, and I would suggest that the significant event was Nicholas Culpeper's English translation of the *Pharmacopoeia.* As the historian Jonathan Sanderson has demonstrated, Culpeper's translation broke the monopoly of the Royal College of Physicians and put herbal and medicinal knowledge back out into the public for the "common good" and for commercial profit.[46] Swiftly following its publication, a torrent of related books were published, including Alethea Howard's *Natura Exenterata* (1655), Elizabeth Grey's *A Choice Manual* (1653), and "Henrietta Maria's" *Queen's Closet Opened* (1655), the latter two going on to become best sellers. Among the first English-language books interested in science, technology, and medicine published under women's names in England, these aristocratic titles opened a door, and many more women of the gentry and the emerging middle class such as Hannah Wolley soon entered the space—and not only women but also men of the middle class. Again, men of the aristocracy were significantly absent until the publication of the *Transactions of the Royal Society.* The close control of the Royal College of Physicians over women's access to knowledge may be one of the factors involved in the exclusion of women from the Royal Society, since 60 percent of the College was elected to the Society between 1663 and 1670.[47]

How did these women from the sixteenth and seventeenth centuries practise science? Like men, they practised in kitchens, stillrooms, and very occasionally laboratories. But unlike men, they worked directly with and for people in their households and local communities. It is a pointed comment that Alethea Howard's book is subtitled *Nature embowelled/ By the most/ Exquisite Anatomisers of Her,* an explicit reference to surgery and the use of one's hands. Andrew Boorde notes in his *Breviary* (1547) that physicians needed astronomy, geometry, and logic for their distinctly hands-off analysis and diagnosis and were expected to talk to their patients but not to touch them. In contrast, a surgeon "need[s] to know the complexion of his pacient, and to consider the age, the weakness, and strength, and diligently to consider if the sickness, sore or impedyment, be perticular by hiselfe: or els that it have any other infyrmyte concurrent with it."[48] A surgeon therefore cannot treat that person in the abstract and has to get his hands dirty. Just so, women were expected to practise surgery and therefore to work with their hands and not their eyes alone.[49] But more important, a surgeon needed to know something about the context of the patient; he, as many of the women discussed above, dealt with the community and the environment out of which an individual came. This was and is a more complex, indeterminate, and rather messy business than that of the distant physician.

In contrast to the men of the Royal Society, the women also worked in specific communal locations: either their family environment, their community, within the confines of the English country house, or among groups of practitioners. They communicated possibly by way of circles of friends, and probably by way of visitors and visits they themselves made. The types of rhetoric appropriate to circles of friends, visitors, extended families, and larger communities generated the communication of and access to a free flow of knowledge documented by the attributions in the many surviving manuscripts, and in some of the printed publications. Communication was also necessary for the testing of suppositions and hypotheses.

The manuscripts of women from the 1530s to 1660 are marked by an increasing awareness of Paracelsian science, with its attendant focus on hypothesis and experiment—Mary Sidney's patronage of the Paracelsian Thomas Mouffet at a time when Paracelsian experiment was not yet widely accepted is interesting to note. But women's manuscripts also consistently combine this awareness of more experimental and hypothetical approaches with a continuance of Galenic attention to the human body as a holistic system and a concentration on the humours theory that preceded physiology. The signal difference between the two methodologies in terms of medicine is that the Galenic assumed that any one patient will need a specific and different treatment from another, while the Paracelsian assumed that one disease will have a remedy common to all people. Women's manuscripts contain evidence of both approaches.

Partridge's *The Widdowes Treasure* lists the remedy "To cause one to pisse" as "approved" and "A precious ointment" is "probatum est"; and many recipes are followed by "It is approved."[50] This Paraclesian testing of general remedies within a book that offered the Galenic strategy of many individual remedies for one illness so that different contexts could be covered was typical. The historian Linda Pollock notes that Mildmay's listing of different cures by context—old, women, young, wounded, etc.—is an explicit attempt at Paracelsian analysis.[51] Howard's *Natura Exenterata* is even more obvious. It offered different cures for a range of ailments, and a whole section of the book was devoted specifically to Paracelsian experiments. The "Address to the Reader" commented that "Method, tis true, may rectifie and informe the reasonable faculty in man, yet be of very little assistance in accidents, whose uncouth causes are not lyable to Rule. . . . They who *do* (though emperically) are to be preferred before those who dispute and talk."[52] In the accounts of remedies, one finds clear notes joining context-specific remark with acknowledgement of "proof": "a Fistula, Canker, or other old soar, which healed the old festered fistula in the brest of Mr Tho. Wood curate of Newington in his dayes. Probatum est," or "Approved . . . by the Lady Capel. 1646," or "proved by her that distilleth the sweets waters at Hampton Court," the last example being doubly important for its implication that women were employed to carry out chemical technology.[53] The Galenic approach to the whole body still anchored the growing Paracelsian experimental understanding of these manuscripts and books in the community. It is interesting that just as the earlier vernacular books explicitly aimed to help the commonweal(th), men writing in this way in the seventeenth century often intended their work also for the common good.[54]

Because these women worked in communal locations, the notion of testing favoured by modern scientists as replicable visual display was not relevant. And because their communication was primarily either oral or tacit, and by informal apprenticeship and observation, there was little need for the formal rhetorical strategies of written proof. The early practice of women differed from that of artisans also working with tacit knowledge and the oral because of its communal location and its noncommercial application. But there were changes as well. By the middle of the seventeenth century, women of the gentry and aristocracy increasingly brought an educated perspective that included astronomy, geometry, and logic to their practice, often learned from the men with whom they worked. Dorothy Moore distinguished, in her short tract on the education of girls, between fantastic secrets like conjuring or manipulation, and real scientific experiment, with the implication that you need testable and replicable knowledge as well as technical skill within your communal setting. The women were bringing together the eye and the hand, linking the two in a social practice that was not class unified, nor gender unified, and which was located in small communities. The democratic access they built used a rhetoric of tacit knowledge and the oral, as well as manuscript, and therefore had a need for contingency, responsiveness, and repetition with variation; they imagined their interventions into nature as requiring a knowledge and practice slow to change.

## Experiment and Representation

The men who were to form the Royal Society were also bringing together disputation and practice, but for different reasons. In itself the incorporation of experiment into abstract disputation would yield little. What was needed was a different understanding of the textuality necessary to the new science. Much earlier than the Royal Society, others interested in opening up the democratic structures of modern science, such as Copernicus, Bacon, and Galileo, recognised a new rhetoric for scientific writing.[55] In late sixteenth-century England, the work of Hugh Platt is possibly the best example of someone attempting to bring the practical experiment of the artisan's commercial secrets into the public domain. Many of his books were balancing acts between protected guild knowledge and new inventions that he wanted to draw to the notice of a wide readership. What Platt lacked was a theory that argued that this knowledge could not be individually owned because it resulted from God's work: it was universal. It took Bacon to hammer out an initial rhetorical strategy. He attempted to recast the topical and situated logic of dialectic and rhetoric of the old disputation as illustrative and expressive of scientific procedure and observation. He recognised the orderly quality of the topics, yet also the way they insist on contingency, repetition with variation, and context. Kenelm Digby also recognised this "order within context" in his comments on translating the popular *Secrets* of Albertus Magnus.[56] Robert Boyle affirmed it throughout his writing.[57]

But there are cognitive problems for the "topics" of classical rhetoric in writing and in print that do not have the same impact on mathematical, syllogistic, or rational logics. The variability and contingency of the topics, such as "quantity" or "degree," bear in a different way on repetition in the oral than in print. In print the relative stability and uniformity of representation does not encourage one to repeat common grounds, or to foreground argumentative premises for different situations. Instead, if you return back over your writing, you do so to check for proofs of earlier premises; so repetition needs to be precise and exactly replicable. To counter this issue in writing as it moves into the duplicating technology of print, a number of results occurred: moving into print culture, one lost the direct relationship between orator and audience. In the written, the reader's immediate response is not to the "character" of the speaker but to the "genre" of the writing, and there was an intense diversification of genre in the sixteenth century. When printed, the marginalian commentary of the manuscript moved partly into the structure of indices and tables of content as well as into the margins of books. Furthermore, the use of manuscript commonplace books, in which writers would collect sayings, examples, and illustrations, and which were often an end in themselves, became instead a highly significant precursor to printed writing.[58]

The topics themselves were gradually delegitimated because they were not reliable enough for maintaining the stable representations ultimately required by modern science. The classical topics suffered by comparison with the non-contextual, non-social logic of mathematics. They were gradually downgraded by a new education system interested in method but not in social context, because that system was socially enclosed and privileged. At the same time the classical topics were co-opted: the categorical topoi of essence, quantity, quality, time, state, relation, place, and active/passive became internalised as phenomena in the new science. They moved from being contingent to becoming *status causae,* or necessary grounds, and were on their way to becoming essentials. By the middle of the seventeenth century, rhetorical inventions, the loci for argument, became the "laws" of science. For example, Boyle's Law—if the temperature is constant, the pressure of a gas is inversely proportional to the volume—is an invention, a proffered probability, but was turned into a universal and became a second-order textuality of modern science.

## The Textuality of Social Practice[59]

The rhetoric found in manuscripts written by women or inscribed for them, such as letters, household recipes, and medical notes, uses the topics as a central device, connecting observation from tacit knowledge with the oral and with the written strategies of the anecdotal and autobiographical, and representing experiment often in diary form: in other words, repeated but contextualised differently by day-to-day change. In addition, most of these manuscripts are collections of recipes from several people. There is no single authorial point of view, but more the kind of editorial voice one might find in a magazine.[60] Because there was no recognised wider public for women's writing until the 1650s, this writing itself usually perceived no need for print, no need for public replicable visual display, and no need for a change in rhetoric or textuality because of a different medium. It may be significant that the published writings implicitly by women prior to the 1650s were midwived by men, and that the first three books ostensibly written by women in the 1650s were published after their deaths and by men. The women themselves may have written manuscripts, but they saw little need for print. After all, those contributing to the Hartlib circle experienced extraordinarily wide circulation of their ideas just by writing letters, or through people writing letters about their practices. But when the topics were delegitimated as reasonable argument by the new rhetoric of early modern science, so also was the rhetoric of the social practice of science by women in local and situated communities.

If the issue for the evolution of modern science in the Royal Society of the seventeenth century was partly one of adapting written and visual rhetoric into a stable technique suitable for representation, the issue for women practising science was quite different. The social practice of science by English women until the end of the nineteenth century, and indeed up until the domestic technology revolution of the mid-twentieth century, was delineated by a range of factors: an ongoing tradition of apprenticeship, communal practice, oral communication, and manuscript record, along with exclusion from a wide public; the denial of access to a more expansive intellectual collaboration; their discouragement from education; and the singular fact that they were not considered citizens even in the world preceding the seventeenth-century social contract and therefore were not included in the new political structure and had no rights to "universal" knowledge. The attempt to reenter formal scientific practice in the nineteenth century was marked in England by a not surprising recasting of household work as "domestic science," but became a dismal failure in its effort to use the encoding textuality of modern science to represent the remnants of communal practice.

Perhaps we should argue that the textuality used by women for their social practice of science was entirely appropriate; we just need to understand what it implied. When Alethea Talbot's *Natura Exenterata* (1655) says that it will offer the "Recreation of Employment" rather than the "Representation of Experiment," we need to learn how to read the structure of the book to appreciate the subtle differences between "re-creation" and "re-presentation"—one of which is that nothing is re-created in the same way, while the purpose of re-presentation is to allow for exact replication of the experiment. Another difference is that re-creation combines the idea of doing with the action of the words, while re-presentation is always an acknowledgeably inadequate mode of expressing the experiment's interaction with the natural world. But another way of looking at this is to argue that women missed their chance, or were deprived of its opportunity, to produce a textuality more appropriate to the social practice of science in a modern world in which social relations were significantly different from those in the late medieval and early modern periods.

Most twenty-first-century scientists recognise that the textuality they work with is inadequate. It is taken as given that words cannot exactly convey the intervention into nature; they can only describe the experiment so that it is replicable, allowing the people repeating to experience the same intervention when they do the experiment. In comparison, communication of technological procedure and information is not attended by these doubts; representation is a necessary matter of accuracy and replicability rather than any attempt to provide the "reality" of nature which science attempts to offer. One of the reasons why writers such as Mary Sidney or Margaret Cavendish, who were not considered "proper scientists," are profoundly interesting to the history of science is because they understood that the texuality of modern science as it evolved is inappropriate to what it is really doing, which is engaging with the physical world rather than representing the experiment. One can argue that each of these women tried to find a more appropriate "re-creational" form for modern science's engagement with nature, but neither attempted to write in a generic form appropriate to the social practice of science. Cavendish was openly aware of the issue, saying, "The truth is, I have somewhat Err'd from good Huswifry, to write Nature's Philosophy, where, had I been prudent, I should have Translated Natural Philosophy into good Huswifry."[61]

Were the social scientific practice of women to be described, far more would be needed to convey the contingencies and variability dependent on social context. And the end would not be replication but varied repetition in a social context. The process would also take more time than the schedules of modern science allow. We still do not have a textuality appropriate to this communication, a textuality that recognises and communicates situated knowledge in science, that is inclusive of a wider public, and that is democratic. This is, of course, the fundamental basis of the feminist critique of the social practice of science and has been ongoing since the seventeenth century.

## Footnote: Another Hypothesis

It seems to me significant that in the throws of the revolution of domestic technology in mid-twentieth-century Britain there was a radical textual shift in writing about one of the areas of women's work in the home: food and cookery. Although there had been earlier examples, works by writers such as Elizabeth David, Jane Grigson, and Alan Davidson introduced the postwar British public at the least to a whole new way of thinking about food and nutrition. They did so not by simply informing people of new products or offering different recipes but by constructing innumerable contexts of possibility for individual readers. Davidson's trilogy of books on fish[62] combines anatomical classification with detailed line drawings and commentary on habitat, all appropriate to a "scientific" book on biology or physiology. At the same time it also contextualises the writer's life and the procedures of his writing and offers a variety of recipes directly connected to the commentary on habitat. Reading such writing is slow work; it asks for commitment. It does not promise replication of an eating experience, because one cannot replicate the habitat. Its knowledge is openly partial, is sensitive to the environment, and invites readers to engage on their own terms. Generically unique, the trilogy offers a far more daringly scientific work than quasi-novelistic accounts by scientists trying to provide context by way of fiction. The signal difference is that writers like Davidson, Grigson, and David, although they became household names, were taking on the writer's risk that honours textuality, knowing that the work may not find a readership, whereas scientists today cannot afford to take that risk and are bound by their contracts to a particular kind of representation that can only become more and more inadequate.[63]

## Notes

1. Samuel Hartlib began to invite correspondence from scholars, writers, philosophers, scientists, travellers, and others from the early 1630s to 1660. He established a scriptorium which copied letters from one individual out to many others and on receiving the replies also copied these back to interested readers. For more background on

the Hartlib papers, see Mark Greengrass, Michael Leslie, and Timothy Raylor, eds., *Samuel Hartlib and Universal Reformation: Studies in Intellectual Communication* (Cambridge: Cambridge University Press, 1994).

2. Boyle's manuscript book of recipes (British Library, Sloane, 1367) has attributions to Barrington, and Hartlib's "Ephemera" has numerous citations to recipes by Barrington, Moore (both as Moore and Dury), and Boyle (as Lady Ranelagh).

3. L. Hunter, "Sisters of the Royal Society: The Circle of Katherine Jones, Lady Ranelagh," in *Women, Science and Medicine, 1500–1700: Mothers and Sisters of the Royal Society,* ed. L. Hunter and S. Hutton (Stroud: Sutton, 1997), 178–97.

4. See R. Maddison, *The Life of the Honourable Robert Boyle* (London: Taylor and Francis, 1969), for a description of Katherine's household.

5. For more information on Dorothy Moore, see L. Hunter, ed., *The Letters of Dorothy Moore* (Aldershot, UK: Ashgate, 2004).

6. See for example, J. Loftis, ed., *The Memoirs of Anne, Lady Halkett and Ann, Lady Fanshawe* (Oxford: Clarendon Press, 1979), 109.

7. A. Searle, ed., *Barrington Family Letters 1628–1632* (London: Office of the Royal Historiographical Society, 1983); the MSS are in the Essex Record Office, D/DBa T40 1/2.

8. F. Harris, "Living in the Neighbourhood of Science: Mary Evelyn, Margaret Cavendish and the Greshamites," in Hunter and Hutton, *Women,* 198–217.

9. Samuel Hartlib, whose papers reside with Sheffield University, for his copying bureau included women's writing, particularly that of Dorothy Moore. These are cited frequently in Hartlib's own notes, his "Ephemera."

10. See Hunter, "Sisters of the Royal Society," 186.

11. The Gresham connection is important since many of the professors went on to become part of the Royal Society, but Gresham College insisted that its professors be single and male, only electing the first female professor in 1996.

12. S. Shapin argues that this occurred because of a need to witness and to consider knowledge claims "in a social setting." See "The House of Experiment in Seventeenth-Century England," *Isis* 79 (1988): 375. Yet despite the useful insight that many scientists first practised at home and then transferred their experiment to the Royal Society, Shapin makes no comment on the women involved nor on the different kinds of social setting that might condition the act of witnessing. See also L. Hunter, *Critiques of Knowing: Situated Textuality in Science, Computing and the Arts* (London: Routledge, 1999), and the discussion in chapter five of repetition as second order textuality.

13. For a useful discussion, see P. Dear, "Narratives, Anecdotes, and Experiments: Turning Experience into Science in the Seventeenth Century," in *The Literary Structure of Scientific Argument: Historical Studies,* ed. P. Dear (Philadelphia: University of Pennsylvania Press, 1991).

14. Of course this is a gradual process toward universals, and the earlier outlook is retained by members of the Royal Society for many years. B. Shapiro argues that Robert Boyle, for example, linked his theory of hypothesis to "man's inability to penetrate the essence of things." See her *Probability and Certainty in Seventeenth-Century England: A Study of the Relationships Between Natural Science, Religion, History, Law, and Literature* (Princeton, NJ: Princeton University Press, 1983), 53.

15. For a discussion of the way universals "implises a co-ordination of particular experiences," see the introduction to D. Gooding, T. Pinch, and S. Schaffer, eds., *The Uses of Experiment: Studies in the Natural Sciences* (Cambridge: Cambridge University Press, 1989), xiv.

16. For detail on books of secrets, see W. Eamon, *Science and the Secrets of Nature: Books of Secrets in Medieval and Early Modern Culture* (Princeton: Princeton University Press, 1994).

17. See Hunter, *Critiques of Knowing,* especially chapter four.

18. As above, see chapter five on the evolution of the liberal social contract. See

also the now classic Carole Pateman, *The Sexual Contract* (Stanford, CA: Stanford University Press, 1988).

19. R. Warnicke, *Women of the English Renaissance and Reformation* (Westport, CT: Greenwood Press, 1983), notes that women were increasingly excluded from guild work from 1450 onward.

20. C. Rawcliffe, *Medicine and Society in Later Medieval England* (Stroud: Sutton, 1995), 186.

21. For example, Lady Anne Butler, the patron of a leper house in Torrington; see N. Orme and M. Webster, *The English Hospital 1070–1570* (London: Yale University Press, 1995).

22. Rawcliffe notes the references to women giving better treatment at home than doctors, referring to the Paston Letters, vol. 1, 218, 291, 628.

23. N. Penny, ed., *Household Account Book of Sarah Fell of Swarthmoor Hall* (Cambridge: Cambridge University Press, 1920).

24. P. Brears and P. Sambrook, *The Country House Kitchen* (Stroud: Sutton, 1998), offers several articles discussing the history of the architecture; see especially the article by Peter Brears.

25. For example, Roger Bacon's book on *Best waters Artyfycialles . . . for the poore sycke* (1530?), the *Treasure of pore men* (1540), and *The glasse of helthe: A great treasure for poore men* (1540). H. Cook, *The Decline of the Old Medical Regime in Stuart London* (London: Cornell University Press, 1986), cites P. Slack's note that there were during this period roughly twenty people for each vernacular medical book in print.

26. Orme and Webster suggest that many hospitals were confused with religious houses and were closed or suppressed in the 1530–45 period; *The English Hospital.*

27. Orme and Webster, *The English Hospital,* 163.

28. See P. Russell, *A History of the Exeter Hospitals, 1170–1948* (Exeter, UK: James Townsend and Sons, 1976), 10.

29. See her treatment for the stone in M. S. Byrne, ed., *The Lisle Letters* (Chicago: University of Chicago Press, 1981), 2:399.

30. D. M. Meads, ed., *Diary of Lady Margaret Hoby, 1599–1605* (Boston: Houghton Mifflin, 1930), attributes this comment to Thomas Bright (255). The Cooke sisters are singled out, alongside royalty such as Elizabeth I and the Princess of Bohemia, as women of wisdom and learning, in Charles Gerbier, *Elogium Hereinum of the Praise of Worthy Women* (London, 1651).

31. L. Hunter, "Women and Domestic Medicine: Lady Experimenters 1570–1620," in Hunter and Hutton, *Women,* 89–107; see also Sister J. Hanlon, "These Be But Women," in *From the Renaissance to the Counter Reformation,* ed. C. H. Carter (New York: Random House, 1965).

32. See Cook, *The Decline of the Old Medical Regime in Stuart London,* who cites F. Poynter, "Patients and Their Ills in Vicary's Time," *Annals of the Royal College of Surgeons* 56 (1975): 142–43; and A. Clark, *The Working Life of Women in the Seventeenth Century* (1919; repr., London: Routledge, 1982), 254–59.

33. W. Harrison, quoted in "The Ladies of Elizabeth's Court," in *Early English Meals and Manners,* ed. F. J. Furnivall (Early English Text Society, OS32, 1868), xc.

34. D. J. H. Clifford, ed., *The Diaries of Lady Ann Clifford* (Stroud: Sutton, 1990); Meads, *Diary of Lady Margaret Hoby;* L. Pollock, *With Faith and Physic: The Life of a Tudor Gentlewoman, Lady Grace Mildmay, 1552–1620* (London: Collins and Brown, 1993). For a hint of Cumberland's activities, see her funeral sermon, British Library, Harl. 6177. See Searle, *Barrington Family Letters,* and the descriptions on 21, 36, 69, 191–92, 230, 232. See T. Taylor, ed., *Letters of the Lady Brilliana Harley* (London: Camden Society, 1853), and comments on 46–47, 53, 128.

35. L. Hunter, "Technical, Domestic and Rhetorical Books, 1557–1695," in *A History of the Book in Britain,* ed. D. F. Mackenzie and J. Barnard (New York: Cambridge University Press, 2003).

36. It was noted as early as John Mirfield's *Breviarium Bartholomei* (1407) that the poor could not depend on the regular life demanded by traditional Galenism that made sensitivity to daily continuity and change a helpful medical factor; F. M. Getz, ed., *Healing and Society in Medieval England* (Madison: University of Wisconsin Press, 1991), xxiii.

37. L. Schleiner, *Tudor and Stuart Women Writers* (Bloomington: Indiana University Press, 1994).

38. For example, see MS 3574 in the Worthing Museum, or the numerous MSS held in the Welcome Library, such as No 213 Acc3988.

39. A. McLean, *Humanism and the Rise of Science in Tudor England* (London: Heinemann, 1972).

40. J. Gerard, *The Herball or Generall Historie of Plantes* (London, 1596).

41. For example, *Natura Exenterata* lists four noblemen, sixteen doctors, and twenty-nine gentlemen; Welcome MS 1340 of the Boyle Family, partly in Katherine Boyle's hand, lists two doctors, two gentlemen, three gentlewomen, and seven ladies.

42. John Partridge's address to the reader in *The Treasurie of Commodious Conceits* (1584) is dedicated to Richard Wistow, an assistant surgeon, with comments indicating that despite some recipes being "hidden secrets," presumably because of repression from the Physicians, he will reveal them here for the public good.

43. Hunter, "Technical, Domestic and Rhetorical Books."

44. This book is *The Ladies Cabinet Opened* (1639), ostensibly by Lord Ruthven, but possibly by his wife, a good friend of the Talbot sisters and Anne Clifford.

45. Hunter, "Sisters of the Royal Society," 182. See also T. Crofton Crocker, editor of the Earl of Warwick's *Autobiography* (London, 1848). Mary Boyle became Lady Warwick. Elizabeth Walker, the wife of Warwick's Chaplain, was said to be an outstanding female practitioner; D. E. Nagy, *Popular Medicine in Seventeenth-Century England* (Bowling Green, OH: Bowling Green State University Popular Press, 1988). On people using Mary Boyle's stillroom as their "shop for chirugery and physic," see S. H. Mendelson, *The Mental World of Stuart Women: Three Studies* (Brighton: Harvester, 1987), 99.

46. J. Sanderson, "Nicholas Culpeper and the Book Trade" (PhD diss., University of Leeds, 1999).

47. K. T. Hoppen, *The Common Scientist in the Seventeenth Century: A Study of the Dublin Philosophical Society, 1683–1708* (London: Batsford, 1972).

48. A. Boorde, *The Breviary of Helthe* (1547).

49. A. L. Wyman, "The Surgeoness: The Female Practitioner of Surgery, 1400–1800," *Medical History* 28 (1984): 22–29.

50. John Partridge, *The Widdowes Treasure* (1585), E8; F5; e.g., G7.

51. Pollock also notes that John Hester's *A Joyfull Jewel* (London, 1579) was one of the earliest books to include Paracelsian recipes, and did so alongside remedies deriving from the humours.

52. The "Address to the Reader" is signed "Philiatros."

53. Alethea Talbot Howard, *Natura Exenterata* (1655), 280, 281, 58. Pollock refers us to Welcome MSS 635, 160 (Ann Brunswick), 363 (Sarah Hughes), and 751 (Elizabeth Sleigh and Felicia Whitfield) for further examples of women integrating the new science with old practice.

54. One example would be John Hester's books, reprinted for many years, for which the 1633 edition of *The Secrets of Physick and Philosophy* (London) claimed to be a necessary contribution to society, without which we see "no comonwealth, no societie to continue long happy" (A5). Kenelm Digby's *Chymical Secrets* (London, 1682) noted that "Tis impious and unchristian to forbear the Publication of those things, which being rendered Publick, well effectually redound to the Advantage and Comfort of miserable Men" (A4), and Robert Boyle's *Medicinal Experiments* (London, 1692–94) provided recipes easily "made serviceable to poor Countrey People." Both these latter books were probably written much earlier in the century.

55. P. Dear argues in "Narratives, Anecdotes, and Experiments" that Boyle used a probabilistic rather than an axiomatic model.

56. K. Digby, *A Treatise of Adhering to God* (London, 1634), A4v; for further commentary, see Hunter, "Sisters of the Royal Society," 191.

57. This is, by the way, one of the reasons why Nicholas Culpeper was also an astrologer, a fact that people today find difficult to understand. Astrology offers a structure for understanding people within a detailed environmental context. Healthcare procedures today take into account when and where you were born, what your family history is, and to what environmental forces your body has been exposed in order to assess present or future risk. Astrology, which is as determinist as genetics (or not), provides an ordered method for presenting all this information. Significantly, Robert Boyle's writing is close to that of his sister, so close that one suspects that effectively they cowrote some texts. For example, the recipe for "Sore Eyes" in her MS Welcome 1340 (Boyle Family) is very similar to Boyle's in his *Medicinal Experiments,* vol. 2 (London, 1693).

58. See A. Blair, "Humanist Methods in Natural Philosophy: The Commonplace Book," *The Journal of the History of Ideas* 53 (1992): 541–51.

59. See Hunter, *Critiques of Knowing,* particularly chapters three, four, and five.

60. The "Address to the Reader" at the start of Alethea Talbot Howard, *Natura Exenterata,* apologises for the recipes being "out of order" because they were contributed by different people, but goes on to say that the publication needed to be hastened for "publick good" and that the table at the back is so accurate that it is "as if" the contributions "had been placed in their order."

61. M. Newcastle, *Philosophical and Physical Opinions* (London, 1663), A2v.

62. A. Davidson, *Mediterranean Seafood* (Harmondsworth: Penguin, 1972); *Seafood of South-East Asia* (Singapore: Federal Publications, 1976); *North Atlantic Seafood* (London: Macmillan, 1979).

63. See Hunter, *Critiques of Knowing,* chapter five.

STEPHAN CLUCAS

# Joanna Stephens's Medicine and the Experimental Philosophy

In the introduction to their groundbreaking collection of essays *Women, Science and Medicine, 1500–1700,* Lynette Hunter and Sarah Hutton have suggested two main reasons why women have been traditionally excluded from the history of early modern science and medicine. Firstly, they point to the fact that the contributions of women to knowledge in the period often took place in the context of what they call "oeconomics" (i.e., within the "primary economic unit of the family within the local community"), an area which has traditionally been overlooked by historians as a site of serious intellectual and technological endeavour. Despite the fact that the "technology with which they worked became a fundamental part of the emerging experimental methodology" of the natural philosophy of the period, women practising within their communities have been persistently ignored by standard histories of science and medicine.[1]

The second reason (closely linked with the aforementioned failure to attend to the "oeconomic" realm) is that the historical narratives of traditional histories of science or medicine have—wittingly or unwittingly—excluded women, together with other marginalised social groups of the period (such as laboratory technicians) from their accounts. In the history of medicine, Hunter and Hutton note, "the learned practitioners and theorists receive notice, but the lowly healers and midwives, the mere 'mechanicks' of medicine are left out of the account."[2] This process, I will argue, begins long before the first histories of science and medicine, in the narratives of physicians and members of the Royal Society in the late seventeenth and early eighteenth century who, in the process of demarcating the institutional and professional boundaries of emergent experimental, medical, and scientific knowledge, constructed models of professional or expert knowledge which exclude women, along with other empirical practitioners, invalidating their practical expertise and the status of their experientially derived knowledge. In this essay I propose to address the changing role of women both as consumers and producers, proponents and critics of experimental philosophy between the mid-seventeenth and the mid-eighteenth centuries in the context of these emergent institutional and professional constructions of knowledge, and to consider the ways in which women's access to experimental knowledge, whether as practitioners or as spectators, was often subject to mediation and framing by male authority.

## Incongruous Ladies: Women and the Distillation of Medicinal Preparations

In the second of two pseudonymous letters in response to Eliza Haywood's *The Female Spectator* in 1745 relating to women and natural philosophy, a male correspondent, "Philo-Naturæ," recalls the ubiquity of women practising domestic medicine in his youth:

> I remember when I was a Boy, the good Ladies were accustomed about this Season . . . to be extremely busy in drying and preserving certain Herbs and Fruits, and distilling others, according to the Nature of the Plants, and the Uses they were intended for, which I found every Woman of Condition then plumed herself very much on a perfect Understanding in.
>
> Wonderful Cures have I seen performed by the Help of Simples prepared in a proper Manner by these good Housewifes. . . . But such Avocations in these politer Days, are beneath the Attention of a fine lady.[3]

The "avocation" of medical practice was, as Philo-Naturæ suggests, an area in which women of the gentry and aristocracy could apply themselves without fear of opprobrium, and I would like to begin by looking at two examples of printed works which reveal the prevalence of this mode of practical experimentation in the seventeenth century, before moving on to look at the increasing contestation of this privileged role in the "politer" world of late-seventeenth-century and eighteenth-century England, through the case of one particular medical practitioner, Mrs. Joanna Stephens.

*A Choice Manuall, or Rare and Select Secrets in Physick and Chyrurgery: Collected and Practised by the right Honourable, the Countesse of Kent, late deceased* contains, as its title suggests, a posthumous collection of medical recipes which had been "practised" by its aristocratic author, Elizabeth Grey.[4] *A Choice Manuall* was published together with another work, which occupies the first half of the book, concerning "most Exquisite Waies of Preseruing, Conseruing, Candying &c," and it seems clear that one of the primary reasons for the acceptability of women engaging in medicinal practice was its close associations with cookery and the kitchen. Many of Grey's remedies, in fact, use the techniques and ingredients of the kitchen in their preparation. A series of "restoratives" are listed, for example, which are essentially mixtures of wine or ale, spices and eggs, and some of the other items of "Physick," such as a "Broth for a Consumption" or a "Cordiall for a Breakfast fasting," seem as much culinary as medicinal. The manner of preparation also seems reminiscent of the kitchen; the ingredients of one recipe are followed by the injunction, "beat them all together . . . [and] put into a stone Jug and set the Jug in a Kettle of water." Other preparations, however, are more reminiscent of the alchemical laboratory, as in the following instructions for the preparation of treacle water: "Take three ounces of *Venice* Triacle, and mingle it in a quart of spirits of wine, set it in horse-dung four or five daies, then distill it in ashes or sand twice over, after take the bottom which is left in the Still, and put to it a piont of spirit of wine, and set it in the dung till the tincture be clean out of it." Many of these preparations call for the use of a "still" or a "Limbeck," and some of the instructions suggest refinements made through practical experience. In preparing a "Soveraign water," for example, we are instructed: "Still it in a Limbeck, and keep the first water by itself, for it is best, then will there come a second water, which is good, but not so good as the first."[5]

There are also descriptions of a number of medical procedures, such as the lancing of plague sores and the application of "Leaden plasters," made by heating a mixture red and white lead, olive oil, and "Spanish Sope." Another notable feature of the recipes for various medicines, including "The Countesse of Kents Pouder, good against all malignant and Pestilent diseases," is their extensive use of apothecary preparations. The Countess's "pouder," for example, contains "Magistery of Pearls, of Crabs eyes, prepared, of white amber prepared, Harts-horn, Magistery of White Corrall, of Lapis contra Yarvam . . . [and] the black tips of the great clawes of *Crabs.*" Other recipes include the use of a range of apothecary preparations such as musk, ambergris, and "oriental Bezar," and common cordial preparations such as "*Carduus* Water," "*Dragon* water," and "rose water."[6] The availability of apothecary preparations, together with the additional charges which might be incurred by having a physician prepare them, made home preparation first a possibility and then a desirability.

Hannah Woolley's *The Ladies Delight: or a Rich Closet of Choice Experiments & Curiosities*[7] contains, as its subtitle indicates, "The Art of Preseruing & Candying" and the "Art of Dressing all sorts of Flesh, Fowl and Fish," but also includes a section entitled "The Ladies Physical Closet," which promises "Excellent Receipts, and Rare Waters for Beautifying the Face and Body." As in Elizabeth Grey's *A Choice Manuall,* the province of the kitchen is extended to accommodate distillation as a kind of "domestic science," and we can see from the engraving facing the title page the physical (and epistemological) proximity of food preparation and distillation, oven and "limbeck."

As in Grey, there is a mixture here of culinary and medicinal preparations, with advice on how to "make a rare Sillibub" or "White Trencher-Plates that may be eaten," jostling together with advice on healing fistulas, or making a "very rare Water . . . [for] deep Consumptions."[8] However, rather than seeking to differentiate the "merely" culinary from the "chymical" in Grey's and Woolley's works, perhaps we should rather consider some of the *continuities* which exist between the two sets of manual and practical procedures. Some of their techniques patently derive from more folkic sources, as in Grey's medicine which uses a mixture of swallow chicks ("feathers, guts and all") fried with herbs and "rotten Strawberry leaves" in "may butter,"[9] or Woolley's recipe for "snail water," which involves the distillation of a mixture of herbs, snails, and "a pint of Earth-Worms slit, and clean washed," whereas others, such as Woolley's preparation of a cosmetic called "lac virginis," suggest the subtle interplay between culinary and chemical manipulation:

> Take of Alumen plumosi half an Ounce, of Champhire one Ounce, of Roach-Allum one ounce and a dram, Sal-gemuni half an ounce, of white frankincense two ounces, Oyl of Tartar one ounce and a half; make all those into most fine Pouder, and mix it with one quart of Rose-water; then set it in the Sun, and let it stand nine dayes . . . then take Littarge of Silver, half a pound, beat it fine and searse it; then boyl it, with one pint of white wine vinegar until one part be consumed . . . then distil it by a Filter or let it run through a thick Jelly bag, then keep it by itself in Glass Vial.[10]

The use of culinary and apothecary ingredients and the use of either domestic or chemical equipment (still or "jelly bag") blurs the distinction between the culinary and the medicinal. Grey too describes processes, which are at once redolent of the apothecary's shop and the kitchen, when she instructs her reader to take the coral, amber, harts horn, and other proprietary ingredients and "beat them all into very fine pouder, and searse them through a fine Lawn searce."[11]

This "domestication" of medicine is also discernable in Gideon Harvey's *The Family Physician* (London, 1676),[12] which gives directions on how to make use of domestic utensils in medicinal preparations at home but also advises his readers on where to buy purpose-made chemical equipment. He suggests, for example, that his readers use "strainers," which could be bought "at the Linnen drapers at six pence the Yard," and "small Presses" from "the Turners at Hosier-lane end next to Smithfield from three shillings to six or seven." The "Glass shop" could also provide equipment: an all-purpose "glass Body" and a "blind head," which could be customised in a manner more redolent of the kitchen than our conception of a laboratory "by applying around the juncture . . . a Hogs Bladder cut into long slips the breadth of two fingers, and dipt in whites of Eggs, beaten thin into Water." Alternatively, the reader could invest in something purpose built; the glass shop could also provide "a glass retort Receiver" and even a "portable Furnace," which was the same price as a "green gallon Head"(2s 6d).[13]

It is clear from these instructions that at last some of London's retail outlets had seen a potential market in chemical equipment for home use. In *The Ladies Delight,* Hannah Woolley stresses her own practical involvement in her "choice Experiments." These recipes are not "taken up on the Credit of Others," she says, "but [I] do Commend them to you from my own Practice." Boasting of her own practical services to the nobility and royalty, she beseeches her readers to "try the Reality of my Endeavours, by your Practice herein." "Be pleased," she says, "to prove any one of them, and do it with diligent care." In her dedication to Lady Mary Wroth, she makes some self-deprecating remarks about the experiments being evidence that she had not "alltogether spent . . . time idlely," and sees that it was a useful diversion for "young Ladies and Gentlewomen" to "passe away their youthfull time, which might otherwise be worse employed." Nonethelesss we should not underestimate this call to practical endeavour, and she expresses the desire that Wroth's patronage will "encourage" the "practice" of women in the kitchen, both in the culinary and the medical sphere. The "Receipts in *Physick* and *Chyrurgery,*" she hoped, would allow her women readers to "be your own Doctors in the cure of the most common diseases incident to the body."[14]

There is a marked difference, however, between Woolley's exhortation to her women readers to "practice" and be their own physicians and Grey's posthumous manual, which has to a large extent been appropriated by its male editor, William Jar, who announces himself in the title of the book as a "Professor of Physick," who has "added several Experiments of the Virtues of Gascon pouder" to Grey's collection. His short narratives are simple records of dispensations and their successful outcomes; the Countess's powder, he says, has been tried by him "many times, with very happy successe . . . upon severall persons by myself, and divers others by my directions." It is clear that the addition of these "experiments" by a "professor" of physic is intended to legitimate the recipes in some way. The dedicatory epistle, while it is addressed to "the most Vertuous and most Noble Lady, *Latitia Popham*" and extolls the "small Manuall" as a "rich Cabinet of knowledge," is in actual fact an oblique act of clientage aimed at the "truely Valiant Colonel *Alexander Popham,*" with whom Jar evidently served during the recent wars, and who had "approved" the excellency of the manual (which Popham describes in military style as "a rich magazine of experience") in "many dangerous exigencies."[15]

Whilst Jar's professional appropriation of Grey's manual seems relatively benign (a fact which probably had more to do with the social rank of his subject than any tolerance of women in medical practice), the toleration of women practitioners began to decline markedly during the last decades of the seventeenth century. This new intolerance, I would argue, must be seen in the wider context of the professionalization of the dispensing of medicines. It is at precisely this time—the end of the seventeenth century and the beginning of the eighteenth century—that a fierce print polemic between apothecaries and physicians gathered momentum. When the physicians began in earnest to lay claim to higher qualifications in the dispensing of medicines on the grounds of their "liberal Education" and their superior knowledge of "*Anatomy* and *Chymistry,*" the "bugbear Apothecary" is increasingly associated with women practitioners, the "Old Woman and [her] *Water Gruell.*"[16]

In this atmosphere of intense male professional rivalry, the women who had been comfortably practicing medicine on aristocratic estates and from their city dwellings were inevitably drawn into the discursive strategies of the embattled male protagonists. It is noticeable in Gideon Harvey's *Family Physician,* for example, that whilst he encourages his readers to prepare their medicines at home, he designates as his primary targets not the physician but the "*Empiricks* and *Little Apothecaries* inhabiting the skirts of the City and Country Villages," the "Herb-women and Druggists" who artificially inflate the prices of their composed medicines. His readers are advised not to encroach on the "great Remedies" of the *"Methodus Medendi"* (such as bleeding and purging), which are the province of the physician. The physician's art, he says, "cannot be understood, unless you have acquired a competent knowledge of the natural Constitution of the parts of the Body of man, their structure, position, relation, and connexion to each other. . . . This knowledge or Science is chiefly gained by frequent Anatomy or Dissection . . . [which is] most certainly the Basis and ground-work of the Art of Physick."[17]

Women's role as medical practitioners, then, whilst it was temporarily an area of scientific knowledge and practice which they could occupy without fear of male opprobrium, was increasingly compromised by the professionalization of dispensing of medicines at the turn of century.

This professional de-territorialisation of women in medicine is quite clear when we look back at *The Female Spectator* of 1745. We left Philo-Naturæ, you will remember, reminiscing about the gifted medicinal ladies of his youth. This leads him to urge *The Female Spectator*'s lady readers to "enquire a little into the [medicinal] Nature" of herbs. He is quite careful, however, to qualify these remarks: "I would not by this be understood to perswade the Ladies to turn Physicians; they may amuse themselves with considering the Nature and use of those Plants . . . without entering into any laborious Study about them."[18]

By 1745 it has become "incongruous" for a "fine Lady to busy herself about Vegetables, either used in the Kitchen or Distillery," although Philo-Naturæ allows that she may concern herself with perfumes and flowers.[19] The kitchen and the distillery are now separate spaces, and those spaces represent distinct provinces. The Lady is now enjoined against becoming her own physician, and a woman's "avocation" has now become a male profession.

## Joanna Stephens's Medicine and the Making of "Truth and Certainty"

The extent to which the professionalization of medicine had become gendered between the second half of the seventeenth century and the first half of the eighteenth century can be seen in the flurry of publishing activity which attended the discovery of a medicine for the stone by Mrs. Joanna Stephens in the late 1730s. Stephens, as one of her promoters writes, was "the daughter of a Gentleman of a good Estate and Family in Berkshire," who had been practising medicine for over twenty years[20] and who appears to have run a small "hospital" in London.[21] In the late 1730s, a remedy for kidney and bladder stones which she had been administering and refining over a number of years brought her a great deal of public attention. With the help of one of her London patients, David Hartley, a fellow of the Royal Society, Stephens promoted her medicine and offered to reveal the composition of her medicine in return for a reward of £5000. At first Hartley strove to raise the money through private subscriptions, but in the period between April 1738 and February 1739 he only succeeded in raising £1387 13s. 6d.[22] In the spring of 1739 Stephens took up the matter in the House of Commons, requesting a sum of £5000 from Parliament in return for making her medicine public, "submitting her Medicines . . . to such Examinations as the House should think fit." On June 16 1739 she handed the recipe for her medicine over to the Archbishop of Canterbury, and it was published in *The London Gazette* the same day. The medicine was subjected to a series of medical trials, which were considered successful by the parliamentary Trustees, and on March 5, 1740, they approved the reward of £5000, which was duly paid by the Exchequer on March 17.[23] What interests me in this apparent success story is the way in which Stephens's medicine becomes the object of male scrutiny and evaluation, and her own practice diminished and occluded, even to the point of disappearance. In the flurry of printed material which preceded and followed Stephens's publication of her recipe, Stephens is a curiously absent presence, even in those texts which were designed to promote her medicine. The reason for this can be glimpsed in the opening paragraph of the act of parliament relating to her medicine:

> Whereas *Joanna Stephens* of the City of *Westminster,* Spinster, hath acquired the knowledge of Medicines, and the Skill of Preparing them, which by a dissolving Power seem capable of removing the cause of the painful Distemper of the Stone, and may be improved, and more successfully applied, when the same shall be discovered to Persons learned in the Science of Physick.[24]

While the parliamentary Trustees were happy to concede that Stephens might possess "knowledge" and "skill," it is quite clear that they do not consider her to be in possession of "Science," and the bill's main purpose seems to be to ensure that the medicine becomes the property of those male "Persons" who claimed particular professional competence in this field: physicians, surgeons, and medical doctors, who (it is assumed) will "improve" the medicine on which Stephens had been working for over twenty years. At every point of the investigation of Stephens's medicine, her practice was subject to approbation by male experimental authority, and finally, I would argue, the medicine itself becomes little more than a site of contestation between male professionals.

Probably the most important contribution to the debate over Stephens's medicine came from her patient and promoter David Hartley, who in an attempt to publicise the medicine while attempts were being made to raise the money from private sources in 1738, published *Ten Cases of Persons who have taken Mrs Stephens's Medicines for the Stone. With an Abstract of some Experiments, tending to Illustrate these Cases.*[25] Hartley's work, as its title suggests, seeks to augment Stephens's long medical experience of the efficacy of the medicine with a series of experiments which claim to "illustrate" the cures. It is not insignificant that whilst not a medical doctor, Hartley nonetheless chose to approach Stephens's medicines via chemical experimentation—that is, via one of the two sciences which had been claimed by physicians as the basis of their scientific expertise. Hartley's initial premise is that Stephens's medicine is able to dissolve the stones in the bladder or kidneys by rendering the patient's urine alkaline. "Since I have been taking Mrs Stephens's medicine," he says, "My Urine has been of a more Urinous Smell than usual, turbid at the time of making, and alkaline, that is, it turns Syrup of Violets green, and ferments with . . . acid Liquors. . . . And as far as I can judge, the Case is the same with all those who take the Medicines."[26]

He then resolves to "try what Effects my Urine, thus render'd alkaline by the Medicines, would have upon Stones taken from a human Body, by digesting them with a gentle Heat." He procured some kidney stones from a surgeon and "digested" them in a variety of liquids: river water, river water mixed with Stephens's medicine, "common urine," and his own "Alkaline Urine." The stones were subjected to a gentle heat for the duration for a week, and then brushed, washed, dried, and reweighed. The stone soaked in alkaline urine had lost most weight over the course of a week. Hartley considered these experimental results to be "strong presumptions in favour of Mrs *Stephens*'s Medicines."[27]

Interestingly, Hartley did not seem troubled by drawing conclusions about the medical effects of Stephens's medicine from experiments done on stones outside of the body. This is because he believed the experimental conditions were analogous to those in the body. The gentle heat applied to the stones was "intended to have been the same as that of Urine in the Bladder." As a fellow of the Royal Society, Hartley naturally assumes the epistemological validity of his chemical experiments, and while he confesses (a little too candidly perhaps) that "There are appearances mentioned in this Abstract, which I cannot account for," he remains confident that "if the Experiments were repeated with more Accuracy, and other new ones tried, it is probable, that many useful Discoveries would arise."[28] In addition to his experimental narratives, Hartley also offers his readings a carefully engraved plate depicting the stones used in the experiments—a good example of the persistence of the "rhetoric of graphics" which John T. Harwood has identified as an important strategy for the experimental philosophers of the seventeenth century.[29] A year later Hartley republished these experiments (tactfully omitting his equivocations about the experimental "appearances") in another work, *A View of the Present Evidence For and Against Mrs Stephens's Medicines,* which now boasted 155 reports of successful cases and some supplementary conclusions and observations on his previous experiments. This work is addressed "To the President and Fellows of the Royal College of Physicians, London," and he submits his evidence, he says, to their "rigorous and impartial Examinations." With these experiments, he tells the physicians, he hopes to prove the "dissolving Power" of the "medicated Urine" so that Stephens will "appear to you in a different Light from common Pretenders to *Nostrums.*" Hartley's experiments then, are seen as epistemological guarantors of the validity of Stephens's medicines, not only having subjected them to the trials of experiment but also having offered them to the scrutiny of the Royal College. With this new audience in mind, Hartley's claims about the solvent powers of the medicine are more considered, and the modes of evidence correspondingly more complex. The question of whether the medicine actually dissolves the stones, he says, is of a "more intricate Nature, . . . As it relates to Facts not so obvious to the Senses; but may be determined . . . by the joint Evidence and mutual Comparison of the Cases, of Experiments made upon the Urine, of Examinations made by the Catheter, and of the Appearances upon opening the Bodies after Death."[30]

The efficacy of Stephens's medicines (or should we say the legitimacy of Hartley's theoretical appropriation of Stephens's medicine?) is now to be established not just by "Cases" (of which Joanna Stephens had doubtless seen many as a practitioner) and chemical experiments, but by surgical examination and anatomy. The appropriation of Stephens's medicine by professional medicine is now complete. Its truth-claims are to be verified by the profession's own theoretical standards, and thus become "Science" rather than simply "Skill," and Stephens's "Nostrum" becomes a "Medicine" to be tried in "private Practice and in the Hospitals, at Home and Abroad."[31] Hartley took particular steps, in fact, to ensure that Stephens's medicine (and his own reputation, as its privileged expositor) travelled "Abroad" by translating his account of Stephens's medicine into Latin and publishing it in Leyden, one of the two major European centers of medical education. In *De Lithontriptico,* Stephens is lauded as *"horum remediorum Inventrix, se multa experientia didicisse."* However, while she is hailed as the "Inventress of these medicines," the rider is significant: Stephens has "taught herself by means of many experiences," with its implication of a merely empirical knowledge.[32] This diminution of Stephens's practice is even clearer in the works of two other experimental promoters of her medicine, another fellow of the Royal Society and clergyman, Stephen Hales, and a medical doctor, John Rutty.

Stephen Hales, an associate of Hartley's who appears to have collaborated closely with him in considering the properties of Stephens's medicine, approaches the problem in a very similar way—that is to say, he experiments in vitro, with various solvents, using stones procured from surgeons. Hales, however, has a slightly different remit from Hartley. He seeks not simply to investigate the medicine's solvent properties, but to improve them:

> Mrs Stephens's Medicines having been found by Experience to have been very effectual in dissolving Stones in the Bladders of several Persons . . . but they being withal very nauseous, by reason of the great Quantity of Soap therein, I thought it proper to inquire wherein the principal Virtue and Efficacy of the Medicines lay, whence haply Means might be found to make them both less nauseous, and more efficacious.[33]

Hales sets himself the task then of isolating what he considers to be the active components of the medicine so that it can be reformulated and "improved" (as the act of parliament had wished). In Hales, as in Hartley, there is an easy assumption of comparability between experimental and physiological conditions, and of the experimental proofs as seen as a theoretical fulfilment of naked experience:

> When to the Event of these Experiments, we add the concuring Evidence of the happy Effects, which the Medicines have had, in making the Urine of those who take them inwardly also a Dissolvent both in and out of the Bladder, then the concurring Testimony of both Methods of Trial, agreeing in the same happy Event, corroborate each other's Evidence, even to the degree of a full Demonstration.[34]

While, like Hartley, Hales stresses the "Analogy there is between the Event of . . . these Experiments, and the Effect which Mrs Stephens's medicines have on her Patients," it is clear that it is the experimental "events" which are given priority in terms of truth-claims, transforming mere empirical "evidence" into a scientific "demonstration." Stephens's part in the process is in fact glossed by Hales as adventitious or providential rather than technical. The discovery of the medicine is described as a "surprizing series of Incidents" by which God revealed the cure to mankind, while later he refers to Stephens having "happily hit on a means" to cure the stone. As a clergyman, Hales is careful to qualify his experimental observations, deferring authority to the medical profession:

> I shall be glad if these Researches prove of any Service in this most important Concern. I make no doubt that considerable Improvements will be made by Physicians, whose proper Province it is, and who are best qualified to find out the means, how to use, these Medicines with Safety, which have so strong a caustick Quality. I have herein been only acting the part of a Naturalist, being excited thereto by the great Importance of the subject.[35]

Stephens's medicine, then, is taken out of her hands and placed in those of men to whose "proper Province" the dispensation (and experimental confirmation) of medicines belongs: the physician and the "naturalist," who are "best qualified" to understand and develop her medicine. In his own work Hales concludes that "the Efficacy of the Medicine lay principally in the Soap-lees and Lime of Egg-shells." He has not "concerned" himself with "examining the burnt Carrot-seeds &c which are of no avail," although he concedes that the honey "is a good Detergent." It is significant that although Hartley and Hales evidently had close contacts with Stephens, and at one point Hales mentions that he is using an "Egg-shell Lime-powder, which was made by Mrs *Stephens* herself,"[36] this is one of the only points in these narratives where Stephens is visible, and nowhere are her own practices mentioned or her comments on Hartley's and Hales's findings recorded.

The largely implicit message of Hales's and Hartley's texts is made explicit in the comments of John Rutty, who in his *An Account of Some New Experiments and Observations on Joanna Stephens's Medicine for the Stone,* presented to the Royal Society in January 1742 and published later the same year, praised the labours of Hartley and Hales in terms which communicate his disapproval of Stephens's role in the process:

> The Medicine as communicated to the Publick, is a Composition operose and troublesome, several Parts of it being of little or no Use, and others plainly calculated to disguise the rest. The Ingredients of which it consists have lately been examined by the ingenious Dr Hales and Dr Hartley, who have with much Judgement rejected the superfluous Parts, and reduced this pompous Medicine to a slacken'd Powder of calcin'd Egg-shells and a solution of Soap. . . . Thus these learned Gentlemen have, in some measure, rescued this Medicine from the Imputation of being merely Empirical; and indeed our whole Faculty must acknowledge ourselves in a particular manner indebted to the accurate Dr Hales for his ingenious Experiments.[37]

Stephens's medicine, then, is in itself "merely Empirical," rendered "accurate" and scientific by Hales's "ingenious experiments." The "pompous Medicine," with the disapproving reference to its attempt at disguising its ingredients, suggests the nostrum of the "Quack," whereas the rational "Judgement" of the "learned Gentlemen" experimenters renders it acceptable to the "Faculty," i.e., the medical fraternity. Rutty's own book, in fact, as its subtitle suggests, also makes use of Stephens's medicine, offering "some Hints for reducing it from an Empirical to a Rational Use." His declared "Design" is to "establish the Credit of these Medicines upon a Basis of Truth and Certainty,"[38] and (as in Hales and Hartley) this necessarily implies an experimental basis perceived to be lacking in Stephens's own practice. The only real innovations involved in Rutty's experiments relate to his trials with a range of alternative solvents (such as limewater and onion, leek, and celery juices) and his use of a variety of different kinds of kidney stones.[39] He also disputes Hales and Hartley's assumption that the medicine is a solvent.[40] While Rutty is more tentative about "reasoning from the Effects of Medicines on Stones out of the Body, to their Effects on the same Stones in the Body," he echoes their emphasis on the "reciprocal Illustration and Confirmation" between the experimental "Effects" and the "Appearances" offered by patients' symptoms. Ironically, after all his slights about Stephens's "pompous Medicine," Rutty's conclusion confirms that it is "the most efficacious Lithontriptick yet known," and he suggests that it is "well worth" a patient's time to "try their Efficacy" rather than "run the Hazard of one of the most terrible and dangerous Operations in surgery."[41]

The experimental trials of Hartley, Hales, and Rutty in their common aim of establishing a "rational" basis for Stephens's medicine, while they are in part a response to her practice as a woman, are also fully consistent with the professional redefinition of medical dispensation which I referred to earlier. For example, if we compare the treatment of Stephens's medicine with Frederick Slare's *Experiments and Observations Upon Oriental and Other Bezoar Stones,*[42] which subjects a well-known apothecary preparation to rigorous experimental scrutiny, we can see many points of contact. Like Hales and Rutty, Slare is keen to give experimental "proofs" of the inactivity of medical ingredients, and is praised by one of his dedicatees, "Dr John Cooke of Bristol," for his "impartial Trial and Condemnation of *Bezoar,* by the Laws of Chymistry, Reason and Experience." His rejection of traditional apothecary ingredients is significantly gendered: "I always thought Bezoar and Pearl fitter for a Lady's Closet or neck, than for a Cordial, unless for a profane and wanton *Cleopatra.* All the precious Gems should be sent back to the Jewellers, Leaf-Gold to the Gilders: Nay, I could for my Part also return the Musk and Amber-Greeze to the Perfumers."[43]

The basic premise of Slare's analysis is that "Substances that contain any considerable Cordial Vertue in them . . . have something volatile or active [in them]," and bezoar's refusal to react with a range of strong solvents is taken as proof that it is unable to "give out any Medical Virtue." In a dedicatory letter, "Dr Robert Griffith, Professor of Physick and Chymistry in Dublin" argues for the superiority of chemical reasonings in medicinal matters:

> There can be no other Way of judging of the Power of Medicine, but either by trying their Effects upon Humane Bodies, which requires long and diligent Observations, or by separating their constituent Parts by Chymical Operations, which has a greater Analogy to the Digestion of the Stomach, than all the mechanical Powers the Mathematicians of late have introduced to explain all the Operations of Nature.[44]

This growing support for the explanatory force of chemistry and the truth-claims associated with chemical experimentation was not, however, satisfactory to Dr. James Parsons, a medical doctor and fellow of the Royal Society, who wrote one of the most severe attacks on Joanna Stephens's medicines and (more particularly) on her experimental advocates. In a long appendix to his *Description of the Human Urinary Bladder,* published in 1742, which considers "Experiments made upon Human Calculi," Parsons attacks the works of Hartley, Hales, and others.

Alluding to David Hartley's Latin exposition of his experiments on the Stephens medicines, Parsons begins by undermining Stephens's claims to be the inventor of the medicine:

> Tho' Mrs Stephens is called *Inventrix horum Remediorum,* the Inventress of these Medicines, and tho' I think there is a strong Insinuation of her having by degrees found out the Calcination of the Egg-shells, and the Addition of Soap to the Powders, in the following Words; yet it shall presently appear, that every Ingredient made use of . . . has been used by very early Physicians . . . since *Avicenna*'s time.[45]

He then quotes Hartley's narrative of Stephens's discovery from the "Supplement" to Hales's *Account,* in which Stephens is described as having "accidentally met with a Receipt" for a stone medicine made with "Egg-shells dried in the Oven," which she then "administered to several Persons":

> After some Trials made with this she began to burn the Egg-shells . . . and it appeared to her that this Powder was more efficacious in proportion as the Egg-shells were more burnt. But finding it often caused great Costiveness, she added a small quantity of Soap occasionally to each Dose, with a View both to prevent this Inconvenience, and also to forward the Dissolution of the Stone. And thus she continued . . . [until] About 12 years ago, she gave her Powder in larger Dozes to one Mr Coxon . . . [who] received a more remarkable cure than any Person before him had done. . . . Upon this she gave the Powder and Soap in still greater Quantities, and found them attended with proportionably greater Success.[46]

This account in Parsons and Hartley gives us the only clear picture we have of the practice of Joanna Stephens, and what we see here is Hartley describing the deductive steps which Stephens took in composing the medicine. It is precisely the "strong Insinuation" of deductive reasoning, however, which Parsons finds objectionable in the account, "for here she is introduced as one rationally improving them," writes Parsons, "and increasing their Virtues by the Calcination of the shells, and the Addition of Soap; as if neither were ever thought of before. It is really surprising that such a Notion could be introduced and credited."[47]

In an effort to dispel the unthinkable notion of a woman practitioner who can "rationally improve" medicines in the same way as her male contemporaries, Parsons takes steps to suggest the derivative nature of Stephens's medicine, suggesting that the recipe "dropt into this Gentlewoman's Hands" from "the *Receipt-books* of Families" rather than being the product of her own practical experience.[48] He gives examples of recipes which are "of very near Relation to those of Mrs Stephens," for example, in Grey's *A Choice Manuall* and another mid-seventeenth-century manual, *Natura extenterata, or Nature unbowelled,*[49] which, although written anonymously, he nonetheless identifies with "The most Illustrious and most excellent Lady Alathea Talbot, &c Countess of Arundel and Surrey," to whom the work is dedicated. The very ubiquity of these kinds of domestic recipes, Parsons argues, makes them suspicious: any number of medicines "met with now-a-days in common Practice," he says, "might be made Secrets of, and their Virtues cry'd up in the same Manner."[50] Stephens's medicine is clearly identified once more with the nostrums of "common Practice," which Hartley, Hales, and Rutty were endeavouring to separate it from.

While Parsons is obviously anxious to diminish his readers' opinion of Stephens's technical proficiency and deprive her of the credit due to her invention, the primary aim of his polemic seems less to condemn Stephens qua women practitioner than to cast aspersions on the experimental philosophy which had appropriated and approved the medicine, by subjecting it to what he considered to be a more rigorous discipline. Parsons's view on experimentation is extremely ambivalent. As a fellow of the Royal Society, he could hardly condemn the ethos of experimentalism outright, and yet his characterisation of it falls far short of praise:

> Altho' Experiments are of great Use in all parts of natural Knowledge, and ought to be vigorously attempted, in order to come at what may conduce to the Benefit of Mankind, and to arrive at the Truth; and altho' it is a laudable Thing to be inclin'd to make Experiments, yet we ought not be so fond of those we make, as to imagine they have proved all we wish for; or to draw such Conclusions from them, as may favour (or skreen any Defect in) what we cry up.[51]

Parsons's book is designed rather to "cry up" the allegedly superior talents of the surgeon, with his firm grasp of "the only Experiment," that is, "opening Bodies after Death." As the book's advertisement of its "Anatomical figures" suggests, the *Description* is first and foremost an anatomical textbook, and Parsons argues fiercely for the experimental priority of anatomy over that of chemistry. Anatomy provides "The only Evidence . . . of Certainty and Truth" in debates concerning the stone. The assumed analogy between experimental and physiological conditions, and reasoning from in vitro to effects *in corpore,* which gave John Rutty temporary cause for concern, becomes a major objection in Parson's counterargument. "When we draw an absolute Parallel between Stones in a Bladder of Kidneys, and Stones in a Glass or Earthern vessel," Parsons argues, "When Soap, Soap-lees, medicated Urine, or nay other Menstruum can remain and act upon a Stone in the Bladder or Kidneys, with the same Constancy and Force . . . as in a Vessel of Earth, Glass &c . . . When we can find saw'd

Stones in the Bladder, and be able to brush them every Day at Pleasure. . . . The Experiments may shew something."[52]

These experiments have "tended greatly to encourage Mankind to an implicite Belief of their being able to dissolve the Stone in the Kidneys and Bladder," Parsons says, but anatomy has shown them to be misleading. His work, in fact, undermines the claims of experimentalism by anatomical means. Parsons takes a number of the cases which Hartley had used to argue in favour of the medicine and records numerous instances where there had been relapses and recurrence of painful symptoms after treatment with the medicine. Not satisfied with merely recording these cases (with, of course, the testimonies of "Persons of undoubted Honour and Veracity"), Parsons took advantage of the demise of some of Stephens's former patients and anatomised them, offering minute anatomical narratives describing the locations of stones in their kidneys and bladders. Answering Hartley's graphic rhetoric with a graphic refutation of his own, Parsons also provided highly realistic representations of the diseased organs in question—for example the diseased bladder of Mr. Gardiner, who was case number XXI in Hartley's *A View of the Present Evidence,* which is accompanied by a detailed case history and a description of the anatomical dissection of Dr. Nourse, a hospital surgeon. Parson's conclusions are not only that Stephens's medicine failed to dissolve the stones as Hartley and Hales believed, but that its caustic nature, by "destroying the Tone" of the bladder, actually rendered it more vulnerable to stones. Whereas Stephens's patients suffered recurrence of the stone, and even the development of new stones, Parsons claimed, "Those who had been cut for the Stone, and are troubled . . . by new Ones generated, are but a few."[53]

Parsons's dispute with the experimentalists makes it clear that Stephens's medicines (and her practice) enter public discourse in the late 1730s and early 1740s as an issue which is framed and appropriated by the professional demands of men. Stephens's medicine became a vehicle of homosocial exchange: in this case an issue over which men could dispute their professional claims to epistemological and therapeutic superiority. There is no doubt that the debate surrounding Stephens's medicines helped foster and promote the careers of the men who were promoting or criticising her medicines. In these bids for career promotion the men were discoursing on Stephens with other men: Hartley addressing the Royal College of Physicians or the European medical fraternity via his Latin "epistolary dissertation," and Hales and Rutty addressing the Royal Society and "the Gentlemen of the [medical] Profession."[54] Hartley was described by one commentator as a "constant Partisan of this Doctoress,"[55] and a letter to Lady Frances Shirley in 1751 shows not only that Hartley continued to champion the medicine in his later life (he died in 1757), but also that in the immediate aftermath of the publication of the recipe in 1739 Hartley and Stephens were involved with a Pall Mall apothecary who made up Stephens's medicine for retail.[56] Stephens's medicines created a genuine stir, both in England and in France, where an examination of the medicine by C. De Geoffroy was presented to the Académie Royale des Sciences on December 23, 1739, and by S. F. Morand in 1742.[57] In 1742 John Rutty, despite his condescending attitude elsewhere in his tract, suggested that "the Success with which the Medicines have been used, deserves the serious Attention of Physicians, and ought to excite them to a careful Examination of their Effects."[58] Notwithstanding her prominence in the medical scene of the 1730s and 1740s, Joanna Stephens, as Arthur Viseltear has noted, has more often been included in the history of quackery than in the history of medicine,[59] a fact which bears witness to the persistent "unremembering" of women's role as medical practitioners in the modern period.

## Notes

1. Lynette Hunter and Sarah Hutton, *Women, Science and Medicine, 1500–1700: Mothers and Sisters of the Royal Society* (Stroud: Sutton, 1997), 2.

2. Ibid., 3.

3. Eliza Haywood, "Philo-Naturæ" to "The Female Spectator," 15 September 1745, *The Female Spectator* (London: T. Gardner, 1745), vol. 4, bk. 19, 33.

4. Elizabeth Grey, *A Choice Manuall, Or Rare and Select Secrets in Physick and Chyrurgery: Collected, and practised by the right Honourable, the Countesse of Kent, late deceased. Whereto are added several Experiments of the Virtues of Gascon Pouder, and Lapis contra Yarvam, by a Professor of physick. As also most Exquisite waies of Preseruing, Conseruing, Candying &c. The Second Edition* (London: Printed by G. D., 1653).

5. Ibid., 17–19, 20–25, 14, 13, 44.

6. Ibid., 105, 108, 12, 184.

7. Hannah Woolley, *The Ladies Delight: or a Rich Closet of Choice Experiments & Curiosities, Containing The Art of Preseruing & Candying both fruits and flowers: Together with The Great Cook; or the Art of Dressing all sorts of Flesh, Fowl and Fish By Hannah Woolley. To which is Added: The Ladies Physical Closet: or, Excellent Receipts, and rare Waters for Beautifying the Face and Body* (London: Printed by T. Milbourn for N. Crouch, 1672).

8. Ibid., 111, 117, 25–26, 285.

9. Grey, *A Choice Manuall,* 63.

10. Woolley, *Ladies Delight,* 66–67, 279–80.

11. Grey, *A Choice Manuall,* 186–87.

12. Gideon Harvey, *The Family Physician, and the House Apothecary, containing I. Medicines against all such Diseases people usually advise with Apothecaries to be cured of. II. Instructions, whereby to prepare at your own Houses all kinds of necessary Medicines that are prepared by Apothecaries, or prescribed by Physicians* [etc.] (London: Printed for T. R., 1676).

13. Ibid., 9, 11–12, 12–13.

14. Woolley, *Ladies Delight,* sig. A2r-v; sig A2v; "To the Honourable and truly Vertuous Lady, ANNE WROTH, Wife to the Right Worshipfull Sir HENRY WROTH," sig.[G11]-[G12]; unnumbered final page, after index, addressed to the "Ladies."

15. Grey, *A Choice Manuall,* "The Epistle," unpaginated sheet between 190 and 191; "To the most Vertuous and most Noble Lady, *Latitia Popham,* Wife of the Honorable and truely Valiant Colonel *Alexander Popham',*" sig. A3r-v.

16. Anonymous, *Reasons why the Apothecary May be suppos'd to understand the Administration of Medicines In the CURE of DISEASES, As well as the PHYSICIAN. In a Letter from an APOTHECARY to a PHYSICIAN* (London: Printed for Luke Stokoe at the Golden Key, 1704), 5, 7, 18.

17. Harvey, *Family Physician,* sig. A2 verso, sig. A5 verso, sig. A8 verso.

18. Haywood, *The Female Spectator,* vol. 2, bk. 10, 4.

19. Ibid., 57.

20. David Hartley, "A Supplement to a Pamphlet intitled, A View of the present Evidence for and against Mrs Stephen's Medicine &c. Being a Collection of some Particulars relating to the Discovery of those Medicines, their Publication, Use and Efficacy, by David Hartley, M.A, F. R. S." in Stephen Hales, *An Account of Some Experiments and Observations on Mrs Stephens's Medicines for dissolving the Stone. Wherein Their Dissolving Power is inquir'd into, and shown* (London: Printed for T. Woodward, 1740), 37–66, esp. 37.

21. John Rutty refers to "Stephens's Hospital" in *An Account of Some New Experiments and Observations on Joanna Stephens's Medicine for the Stone: With some Hints for reducing it from an Empirical to a Rational Use [. . .] Presented to the Royal Society Jan 14 1741–2 By John Rutty, MD* (London: Printed for R. Manby, 1742), 13.

22. David Hartley, "Proposals for making Mrs Stephens's Medicines Public," in *A View of the Present Evidence For and Against Mrs Stephens's Medicines, As a Solvent for the Stone. Containing a Hundred and Fifty-Five Cases with some Experiments and Observations* (London: Printed for S. Harding, J. Robinson, and J. Roberts, 1739), 175–82. The "List of Contributions [. . .] from April 11th 1738, to February 24th 1738–9" (177–82) includes several large contributions from the nobility (£50 each from the Earls of Pembroke and Scarborough, and £100 from the Earl of Godolphin) and numerous smaller contributions from the gentry, including Hartley himself, who invested £5 5s. in the venture.

23. Hartley, "Supplement," 38–39.

24. "An Act for providing a Reward to JOANNA STEPHENS, upon a proper Discovery to be made by her, for the Use of the Publick, of the Medicines prepared by her for the Cure of the Stone." The act is reproduced in toto in Hartley, "Supplement," 39–42. The opening paragraph reproduced here is to be found on 39–40.

25. David Hartley, *Ten Cases of Persons who have taken Mrs Stephens's Medicines for the Stone. With an Abstract of some Experiments, tending to Illustrate these Cases* (London: Printed for S. Harding and J. Roberts, 1738).

26. Ibid., 28.

27. Ibid., 29–30.

28. Ibid., 29, 38.

29. John T. Harwood, "Rhetoric and Graphics in *Micrographia,*" in *Robert Hooke: New Studies,* ed. Michael Hunter and Simon Schaffer (Woodbridge: Boydell Press, 1989), 119–47.

30. Hartley, *View,* iii, iv–v, 1–2.

31. Ibid., v.

32. David Hartley, *De Lithontriptico a Joanna Stephens nuper invento Dissertatio Epistolaris auctore Davide Hartley, A.M., R. S. S.* (Leyden: Johann and Hermann Verbeek, 1741), 5.

33. Hales, *Account,* 1.

34. Ibid., 26.

35. Ibid., 13, 30, 33.

36. Ibid., 33, 28.

37. Rutty, *New Experiments,* ii.

38. Ibid., title page, vi.

39. Ibid., 15–18, 28–39, iii ("I thought the Importance of the Subject required the like Experiments to be repeated and diversified on a greater variety of Calculi").

40. Ibid., iv ("The several ingredients of his Medicine [. . .]. do only corrode and precipitate the Parts of the Stones, and therefore in Strictness should rather seem to deserve the Appellation of Lithontripticks then Solvents").

41. Ibid., iv, 37–38.

42. Frederick Slare, *Experiments and Observations Upon Oriental and Other Bezoar Stones [. . .] Dedicated to the Royal Society* (London: Printed for Tim Goodwin, 1715).

43. Ibid., xvi, xvii.

44. Ibid., 5, 13, x.

45. James Parsons, M.D., F.R.S, *A Description of the Human Urinary Bladder, and Parts belonging to it: with Anatomical Figures [. . .]. To which are added Animadversions on Lithontriptic Medicines, Particularly those of Mrs Stephens; and an Account of the Dissections of some Bodies of Persons who died after the Use of them* (London: Printed for J. Brindley, 1742), 134.

46. Ibid., 135–36. Cf. Hartley, "Supplement," 37–38.

47. Parsons, *Description,* 136.

48. Ibid., 151.

49. Anonymous, *Natura Exenterata, or nature unbowelled by the most exquisite anatomizers of her: wherein are contained her choicest secrets digested into receipts [. . .] in the art of medicine [. . .] Whereunto are annexed many rare, hitherto unimparted inventions, etc.* (London: Printed for H. Twiford, G. Bedell, and N. Ekins, 1655). The dedicatory epistle is signed "Philiatros."

50. Parsons, *Description,* 151–52, 137.

51. Ibid., 113.

52. Ibid., 114, 241, 114–15.

53. Ibid., 114, 241, 233–40 (cf. Hartley, *View,* 27–28), 249, 243.

54. Rutty, *New Experiments,* 39.

55. Anonymous, *An Address to the Gentleman of the Faculty of Physick; in Answer to A Letter, relating to Mrs Stephens Medicines, inserted in the Daily Advertiser on the 22d of August last; and the Queries therein propounded fully consider'd* (London: Printed for W. Meadows, J. Clarke, and R. Partington, 1739), 8.

56. David Hartley to "The Right Honble The Lady Frances Shirley, in Dover Street London," dated "Bath, Apr[il] 27 1751," British Library, Additional MS 61667, ff. 144–45. See esp. f. 145 r.: "If the Person for whom your Ladyship intends it sh[al]l want any farther Information ab[ou]t the Manner of making it, he may consult Mr Roberts an Apothecary in Pall-Mall, at the Golden Mortar, who made the Medicines after Mrs Stephens's Direction, when we published them in the Gazette, & who understands their Nature much better than any Body I know."

57. Arthur J. Viseltear, "Joanna Stephens and the Eighteenth Century Lithontriptics: A Misplaced Chapter in the History of Therapeutics," *Bulletin of the History of Medicine* 42 (1968): 199–220, esp. 209–10.

58. Rutty, *New Experiments,* ii.

59. Viseltear, "Joanna Stephens," 200.

MONIKA MOMMERTZ

Translated by Julia Baker

# The Invisible Economy of Science

## *A New Approach to the History of Gender and Astronomy at the Eighteenth-Century Berlin Academy of Sciences*

"At lunch time, shortly after eleven o'clock, I began my observation. I was working at the big quadrant [upstairs], Christinchen, downstairs, with the small quadrant. By knocking I indicated to her that I had measured the altitude. We had intermittent sunshine with clouds."[1] Short descriptions such as this one once constituted the results of observations of astronomical phenomena in early modern Europe. During the seventeenth and eighteenth centuries, these diaries of observation were used to prepare publications—ranging from calendars to scientific papers—or they served as the basis for the international exchange among people at many levels of education and of all social classes.

There is another way to read the quotation: as a description of a typical working situation in the then-developing science of astronomy. Thus, for readers of the twenty-first century, the world of the working reality of early science unfolds. This might seem strange in some respects. In the first decades of the eighteenth century, the fields and epistemological contours of the new sciences of nature were not yet defined, and strange professional constellations were no exception. Doctors could also be philosophers and craftsmen. Lawyers specialized in botany and philology. Members of the middle classes and even farmers dedicated themselves to the fields of meteorology as well as astronomy, and they did so at the scientific level of the time.

During the nineteenth century, the sciences became institutionalized and embodied at the universities. With the specialization and differentiation of fields, science soon distanced itself from the earlier forms of scientific work and methods. This led to a devaluation of non-institutionalized methods, which continued far into the twentieth century. "Unprofessionalism" and "dilettantism" were among the most persistent allegations against the early practitioners of the sciences. In fact, this was a one-dimensional view, which affected the history of women in science most fatally, since by devaluing all aspects which appeared not to promote a narrowly defined concept of "progress," women's activities were more often than not ignored and overlooked. Only in recent decades, when the traditional historiography (based on images of male pioneers of early science) has been criticized more and more, did researchers begin to find that women also took part in and influenced the development of modern science.

This paper deals with three aspects of this change in the practice and history of science. First, it adds to our knowledge of women who worked in a scientific context and who had their own personal scientific interests and views. The quotation cited at the beginning of this essay was made by Maria Margaretha Winkelmann-Kirch, who among her contemporaries was known as an astronomer. As it also shows, she had educated her thirteen-year-old daughter, Christine, to work on the quadrants. The paper not only will describe Maria Margaretha Winkelmann-Kirch's work but also will present the ways in which her daughters contributed to the Berlin Academy of Sciences.[2] For, as the quotation illustrates, Maria Margaretha and her daughters, Christine and Maria, together carried out observations for decades. There was also a third daughter, Theodora (Gottfried Kirch's daughter from his first marriage), who participated in the observations. Unlike the male members, the women of the Winkelmann-Kirch family were never accepted in the circles of the Berlin Academy of Sciences. However, it will be shown that these women nevertheless performed very important activities for the Academy throughout the eighteenth century.

The history of their working lives and accomplishments has another purpose as well; it broadens our understanding of how gender affected the institutionalization of the first "natural sciences." The coming into being and expansion of the Prussian monarchy led to the foundation of the Berlin Academy of Sciences in 1700. Past research on the role of women in this era by scholars such as Londa Schiebinger has been marked by the idea that the founding of the large scientific academies was a "turning point for women in science" in that it meant the exclusion of women—such as Maria Margaretha Winkelmann-Kirch in Berlin—and led to their ultimate ejection from institutions where *organized* research took place.[3] However, the unexpected amount and significance of continued astronomical activity by a number of generations of women in the Winkelmann-Kirch family for the "academic" sciences casts doubt on this analysis.

In this essay, therefore, I want to demonstrate how, for over a century, old and new forms of scientific work were combined at the Berlin Academy. In this respect, a particular form of organization of scientific work merits special attention. The Winkelmann-Kirch family, and particularly its female members, had been working in the field of astronomy before the Academy came into existence and continued working for the Academy for the next seventy years. In fact, the "household" as a way of life, economy, and production maintained its importance as a place where scientific work was performed. Even after "official" observation moved to the Academy, what I will refer to as the culture of the astronomical household functioned as a long-term efficient resource for the institution.

Third, this essay documents why and in what way we can refer to this situation as "an invisible economy of the sciences." Not only does this term describe how the achievements of these women were fundamental to the continuance of the Berlin Academy in the eighteenth century, but also how the Academy consciously tried to render these achievements invisible. Though perhaps unacknowledged, the women's activities constituted one of the formative working methods of the early Academy. Embedded in the structure of relation and organization, the hierarchies, values, and positions of the household, astronomical observation occurred at first within the scope of a *self-contained* working system. In the course of the eighteenth century, the astronomical household was connected to the emerging institutional working system of the Academy. During this evolutionary period, both working systems were systematically intertwined. Their interrelations followed their own patterns. They underwent changes, but were always functional and at the same time asymmetrical. The work methods in the astronomical household, circumscribed by the category of gender, continued to comply with the needs and requirements of the institution and, thus, were determined by it. Although the expulsion of women from official affiliation to scientific academies did occur during these years, this essay attempts to contrast Schiebinger's abrupt "turning point model" with a "phase model." By showing the most important phases and turning points in the collaboration within the different household configurations of the Winkelmann-Kirch family, the achievements of the women can be made visible, and their activities for the institution reevaluated. The "shadow economy" of their astronomical production demonstrates that in women's working relations with the Academy, their contributions—albeit made invisible and controlled by the Academy—proved to be extremely profitable for the latter. In the context of research on gender and science, the concept of the "household" as one of the working systems in early modern science presents a promising new analytical approach.

## The Working Reality of Science in Early Modern Europe

European women's participation in science in early modern times is dependent on the close interrelation between the surrounding circumstances of life and science. Thus, the historical search for women in the university environment, which was frequented mostly by men until the nineteenth century, proved to be largely unsuccessful.[4] In other social areas, which do not seem to be scientific at first glance, the search was more successful. Papers on the history of women or gender point out aristocratic women who received their (mathematical-philosophical) education at the court and discussed their findings with private tutors. Female members of the middle classes were taught by their fathers and husbands following the humanistic ideal of the *"puella docta."* They created their own ideals of female erudition and discussed their findings in frequent newsletters. Historians also pointed out that there were women who participated in various ways in creating new learning. They helped their husbands in their scientific work; female lecturers and professors used the political conflicts and the plotting and scheming of their male contemporaries to obtain places within the academic world; women dominated "family firms"; women created their own communities to discuss their scientific findings; and, finally, European women traveled and collected, and participated in colonial discoveries.[5]

When reading such studies, it is apparent that all of them share one aspect, which is only mentioned incidentally in each individual study. For women in the early modern scientific world, the household played a major role in many respects. Through it, women gained access to education, or it functioned as the necessary condition of their research. In recent studies, which deal more with the history of science, without the analysis of gender, one notes that the terms "household," "family," "domestic realm," "the house," and "private sphere" are still in evidence, but have become interchangeable. They are no longer used to shed light on the participation of women in science.[6] In addition, unlike the relation between "science" and "the public," the household as a site of scientific production has not been researched in a systematic or detailed way. Instead, the prevalent schematization of "private" and "public" seems to cover up women's scientific contributions. Apart from a few exceptions, the history of astronomy and the Academy—particularly the historiography of the Berlin Brandenburg Academy of Sciences—does not deal with male- or female-headed households of scientists who worked for these institutions. As far as the official membership is concerned, the decisive criteria for the choice of subject, studies, and achievements of women or other gender aspects are overlooked, if not ignored. As this essay will show, however, many forms that were decisive for the development of "public" and "collective" cooperation were implemented in the framework of household culture.

Despite the household's numerous forms (e.g. bourgeois, aristocratic, craftsmen, etc.) and differences in space, time, and religion, the early modern western European household was *the* basic organizational form of early modern society.[7] It constituted a social unity, to which the community as statutory corporation, such as guilds and universities, could relate. The household under the authority of its "head" was also seen as the basic production and reproduction unit; it served as a place where people were protected and fed. It often functioned by involving generations. The household constituted the primary work force of female and male members of the family. "To household" also meant to provide work-related socialization and education of children and adolescents.[8] Work strategies of household members in most parts of the population were based on the maintenance of shared material and cultural goods and did not focus on the individual—though gender, space, and class were still important. Through kinship, trade, and loan relationships, as well as neighborhood ties, the household formed a network with other households in mutual dependency. Depending on the type of the household, different forms of sociability were typical. An organizational form among household members that was based on mutual dependency was only possible as long as they worked within a hierarchical structure based on gender, age, and class. A king's court, a businessman's shop, the workshop of a craftsman, a convent, etc. could be referred to as a "household" and function along similar principles.

Against this background of the fundamental social meaning of "household"—as a form of "culture" with its many possible specifications—I would like to point out how such elements corresponded with the life and work methods of the Winkelmann-Kirch family in its various phases, and its relationship with astronomy over time. Whether our case study is exemplary or not will depend on further research. As a very important social and cultural institution of early modern times, the "household" describes both the structural place of gender, as well as a significant location outside of the university, which contributed to the development of modern sciences. In this essay, particular attention has to be given to the gender specific models of cooperation within the household, such as the "male-headed household." Furthermore, other basically hierarchical forms of cooperation as contextual conditions of scientific work (involving members of the same/opposite sex), such as married couples, father-son or housewife-servant relationships, and relationships among siblings, must also be considered.[9] The example of the Winkelmann-Kirch family shows the cultural framework in which methods of scientific production took shape. Emphasis is also given to the working system of the household and how the relationship between men and women in science was affected after the Berlin Academy was founded.[10]

## Models of the Household Production of Science: From Apprenticeship to Master Astronomers

When in 1692 Maria Margaretha Winkelmann married Gottfried Kirch, he had already raised eight children with his first wife Maria Lange, who had died in 1690. Kirch's life with his first wife can be traced back to Jena, Danzig, Königsberg, Langgrün (near Lobenstein in Thüringen), Coburg, Leipzig, and to his birthplace, Guben. Since universities at that time did not offer practical astronomical methods, he had acquired them under the supervision of the well-known astronomers Eberhard Weigel in Jena and

Johannes Hevelius in Danzig. While working at the observatory of the wealthy Hevelius, an educated lawyer and brewer, Gottfried Kirch might have witnessed his mentor's collaboration with wife, Elisabeth Koopmann. This experience might have influenced Kirch's decision to choose a partner educated in astronomy for his second wife.

During his first marriage, calendar production had provided an unstable and therefore uncertain source of income. The family had to remain geographically flexible to find education, income, and patronage. Gottfried's family therefore followed him to his numerous residences.[11] During this first period, Gottfried and Maria Lange's daughter Theodora worked with Gottfried on his astronomical observations.[12] She received her education, like her brothers, from her father. Unlike her brothers, she was not accepted at the university. The observation diaries of later years document that Theodora was skilled in astronomy and that she participated in observations for the Academy. On the other hand, there is no evidence that Theodora's mother, Maria (Gottfried's first wife), participated. For this period prior to the work for the Academy, it can be noted that members in this astronomical household already worked together and were educated in it.

Maria Margaretha Winkelmann had lived and worked in different places before she met and married Gottfried Kirch. Her first teacher was her father, Matthias Winkelmann, a Lutheran pastor in a small town who is said to have encouraged and supported her interest in astronomy.[13] Maria Margaretha became an orphan at the age of thirteen, and the inspector of the orphanage in Halle (her father's successor as pastor and later her brother-in-law), Justinus Toellner, provided her with a broad education, which included mathematics and Latin. She entered the household of the wealthy farmer Arnold in Sommerfeld (near Leipzig), presumably as a maid. Arnold belonged to a group of amateur scientists who carried out astronomical research in their spare time. His letters even appeared in correspondences among members of the *res publica litteraria* (Republic of Letters).[14] Maria Margaretha Winkelmann's position—pastor's daughter and maid—in the household was certainly ambivalent. In Arnold's house, she learned the basics of astronomy, meteorology, and observation techniques. The young woman took notes of her observations and thus began what became the extensive meteorological research of the Winkelmann-Kirch family in later times.[15] Thus, education and work, socialization and first contacts with scientific research were combined in this household.

The marriage of Maria Margaretha Winkelmann and Gottfried Kirch in 1692 marked the coming-into-being of a household that focused on collaboration in the research of astronomy and astrology. Winkelmann and Kirch were not only united by their scientific interests. Most likely, their religious affinity also brought them together. They believed in pietism, a fact which has been disregarded by many historians for a long time.[16] In Leipzig, one of the most important sites of these Protestant reformers, Kirch (and most likely also his wife) belonged to a radical group, which had been prohibited and more than once prosecuted by the government in Dresden. After spectacular arrests of some of their fellow Protestants, the couple had to leave the city and moved to Kirch's hometown, Guben, where in 1696 Christfried, a little later Christine, and shortly afterwards Margaretha were born.[17] There might have been a third daughter, Johanna, but no further information exists about her.[18]

As many letters document, Maria Margaretha's interest in astronomy and astrology had much to do with her religiosity. She, like others, was convinced that God influenced the weather and other meteorological phenomena. At the time, this was the widely accepted common religious belief among educated people, although astrology in its narrow sense had often been under attack.[19] Pietism might have played a major role in her development. Like other female believers with whom she apparently was in contact,[20] Maria Margaretha benefited from the movement's focus on the immediate relationship of the believers to God. This focus provided women with enhanced competence in religious and secular affairs. In this religious movement, it was not uncommon for women—like Maria Margaretha—to work independently in groups and towns and even to become preachers.[21] In our context, the religious orientation is important because pietism influenced Maria Margaretha's self-confidence and shaped the relationship of the couple. Their working relationship can thus be viewed in a new perspective. Right from the beginning of the marriage, the Winkelmann-Kirchs worked closely together, a "working married couple," inspired and motivated by their shared religious belief. It can be noted as something extraordinary that the Winkelmann-Kirchs shared a working relationship in which the wife independently carried out her astronomical studies and published academic papers in her own name.[22] Her husband acknowledged her work; he wrote down her results in his observation diaries on a regular basis. This relationship, which appears equal from the outside, is not typical for a working couple of that period and can best be explained by their radical religious background.

Until 1700 the family lived in different places.[23] The household provided a degree of independence from patrons and institutions, but was also burdened with a certain insecurity. The family produced astronomical-astrological calendars, which were used by all classes in the population and which therefore sold well. For this purpose, calculations had to be carried out, additional texts and illustrations had to be produced, and the calendars had to be sold to booksellers and publishers in cities like Breslau, Nuremberg, and Danzig. At the same time, Gottfried and, far less extensively, Maria Margaretha were in correspondence with other scholars and self-educated astronomers. Maria Margaretha seems to have been very much involved in the production. She also encouraged her daughters to learn astronomy and calendar making. In a letter, one of Gottfried's sons, Gottlieb, mentions how unimportant she considered other—more or less thought of as typically female—household duties. In a letter to his father, he complains that she did not educate his sister Theodora in important ways. She did not teach her spinning, sewing, cooking, or knitting.[24] Gottlieb does not mention that Maria Margaretha taught Theodora, and her other children, everything she knew about astronomy. At the same time, his letter suggests that many activities within the household were performed by servants. Even so, during this period, money was scarce. The worries continued but could be compensated for by a specific but very flexible organization in the astronomical household. The meaning of providing for each other within the household becomes apparent here. The family members, including the children, as they grew older, were all involved in the working process. If one was sick or somebody moved away, somebody else took his or her part.

## An Astronomical Household with an Academic Connection: The First Years at the Academy

The founding of the Königliche Societät der Wissenschaften (Royal Academy of Sciences) in Berlin brought about important changes. Shortly before the turn of the century, the philosopher-physicist Wilhelm Leibniz mentioned the Winkelmann-Kirchs to the elector Frederick, who became King Frederick I in 1701. After Leibniz's recommendation was accepted, a new phase of the Winkelmann-Kirch astronomical household began. The family moved to Berlin, and, with Gottfried Kirch as official astronomer, the household received its income from the Academy. Astronomy, as one of the queen's favorite subjects, was given special consideration in Berlin, which had become the seat of royal power. In his position as astronomer, Gottfried became directly committed to the venerable institution. For the Academy, astronomy became a representative and lucrative branch of science. With the acceptance of the Gregorian calendar, the Protestant estates had finally given up their opposition to the mathematically and astronomically more correct measurement of time. This opposition had lasted for one hundred years because the new methods had been labeled Popish and Catholic. Following Leibniz's advice, Frederick I took this opportunity to bring the Prussian calculation of time and calendar making, which had been a source of income for various calendar makers, editors, printers, and merchants, under royal control. He gave "his" Academy the monopoly of calendar making and thus arranged the major source of its funding until the turn of the nineteenth century. The Winkelmann-Kirchs proved to be vital to this project.

As far as household production was concerned, working for the Academy did not bring about many changes. On the one hand, this was because there was not enough space at the Academy to perform observations. Like other major European academies, the Berlin Academy should have had an observatory, but it had yet to be constructed. Therefore, a major part of the work still took place in private homes—particularly at the observatory of the patron of the family, Bernhard von Krosigk.[25] The future working place of the astronomer, the "royal observatory," was partly in use from 1706 onwards, was ready for occupancy in 1709, and finally was inaugurated in 1711. Even so, although some external conditions had changed, the family continued their astronomical work from home. Gottfried's diaries, which had to be handed over to the council at the end of each year, along with Maria's, document how the family continued to work as they had before they moved to Berlin.[26] The research still took place in the same household environment. The family made observations from the windows, the yard, the garden, and the attic. The astronomical work in the narrow sense was embedded in the everyday rhythm of family life. Sometimes they had to stop their studies as "two batches of laundry were hung in the attic," or simply because somebody had lost the key.[27]

The division of labor was dependent on the spatial coordinates of the household. In order to determine changes of an astronomical phenomenon, the family had to record when those changes started. In that case, one of the family members ran to the clock in the living room, noted the position of the hands, and reported the exact time back to the observing person. Depending on where they observed, they gave each other signals by knocking. Basically, the family made observations day and night, and Gottfried and Maria took turns, as observing could last from the early hours before midnight to the later evening hours, or even the entire night. One single person could not have covered these long periods of time. Neither could one person have coordinated the information between clock and telescope.

Both Gottfried's and Maria Margaretha's diaries report that visitors helped with the observation, often for hours or even days. They were well-known and unknown figures, members of the Academy, local members of the educated classes, or visitors from abroad. The Winkelmann-Kirchs also had useful contacts among people who did not belong to the educated classes.[28] A conjointly organized household also included one's own business and private relations, which were established to a great extent in letter exchanges on astronomical topics. Although Gottfried apparently had a more extensive correspondence, the members of his family, and especially Maria Margaretha with her own correspondents as well as her astronomical contributions, are very much present in Gottfried's writings. The particular time economy of the family allowed its members flexibility and even replacement in case of illness. This fact proved to be practical in regards to the Academy. Due to his age and his deteriorating health, Gottfried was not able to work for long periods of time. This was most alarming as the calendar had to be completed at a certain time. In that case, Maria Margaretha and the children took over. She was responsible for providing the calendar scripts on time.

Between the period before and after 1700, a transition of the representational level has to be considered. From the beginning of the relationship between this astronomical household and the Academy, the husband was the only recognized mediator between both working systems. The new institution intensified the gender hierarchy of the working couple, which had been less apparent in the independent household. According to the Academy, the control of the work to be done was the duty of the person carrying the title of astronomer, a title which—in this case—was gendered male and became identical with the head of the household.

## The Astronomical Household and Female Supervision: Between Dependence on and Independence from the Academy

After the astronomical household had become settled, it continued to function in the described fashion. It seems to have been Gottfried Kirch's death in 1710 that introduced major changes to the working division of the family. After his death, Johann Heinrich Hoffmann was elected to succeed Kirch as astronomer and was made a member of the Academy. He officially had to perform the entire workload, which until then had been organized and performed by the Winkelmann-Kirch family. As Londa Schiebinger has pointed out, Maria Margaretha petitioned to be allowed to carry on her husband's calendar production. The Academy council rejected all her applications in this matter.[29] Indeed, Gottfried's former assistant, now the new astronomer, could not cope without his rival's experienced knowledge, a fact which he tried to hide. Hoffmann's lack of skill was repeatedly criticized and discussed by the council. It is very likely that the missing skills and labor of the Winkelmann-Kirch family were noticed immediately. The skills and knowledge acquired in the astronomical household over the years, could obviously not be easily transferred or acquired elsewhere.

Despite arguments with the Academy, Maria Margaretha at first remained in the apartment appointed to the astronomer by its council. Soon she began using an observatory provided by Baron Krosigk, where she and her children lived together until 1714.[30] Not much is known about the following period. The family members lived separated from each other for some time. Christfried moved to Nuremberg for some months, while his mother and sisters stayed in Berlin. The family still tried to provide for each other, and the women led their own household. Without the Academy's salary, one has to imagine the family's situation more precarious than ever. Astronomy remained part of the life of the Winkelmann-Kirch family as they continued their research. According to Maria Margaretha's observation diaries, her daughters, Christine and Maria, constantly assisted her.[31] In addition, this female household continued the production and distribution of calendars—however, due to the royal Academy's monopoly on calendars, they had to limit production to a great extent and could only sell outside of the Academy's territory, i.e., Prussia.

This phase of insecurity was followed by social and financial promotion. In 1716, shortly after Hoffmann's death, Christfried Kirch was hired by the Berlin Academy of Sciences. He shared the position of "observator" with Johann Wilhelm Wagner, a mathematics teacher from Berlin. The Academy did not reappoint an astronomer for many years to come. Meanwhile, Wilhelm I, known for his skepticism about the sciences, had become Prussian Emperor. He was primarily interested in the Academy's achievements in regards to manufacturing and the military. With Wilhelm's strict frugality, the Academy could not afford to hire an astronomer. The two observators had to fulfill the needs. When Wagner left Berlin in 1720, Christfried, who had been an ordinary member of the Academy, was—as a result of his own urging—finally promoted to the position of "astronomer."[32]

There is considerably less information available about the women's work during this period than during Gottfried's lifetime. Although Christfried's numerous observation diaries continue where his father's ended, Christfried mentions hardly anything about the cooperation among family members. At first glance, his mother and sisters seem to be absent.[33] On closer examination, indirect signs of their active participation can be noticed. In fact, the house remains a place of observation until Christfried's death in 1740. "I observed at home," he regularly notes in his books, which he presented to the council. Certain parts of the work process could not have been carried out by one individual. In some places there are also abbreviations of one of his sisters' names.[34] Planned and in an organized way, the sisters, or at least one of them, were involved. They may even have carried out the entire astronomical work at home.

Only Christfried could move between the two work places, the household and the royal observatory at the Academy. After their mother's expulsion from the representative rooms of the institution, the sisters were also denied access. They might have broken this rule once, but Christfried should not or did not want to make this fact known. The management at the Academy criticized his working at home, without mentioning the sisters' participation. Christfried in turn often complained about the "useless" assistants who were provided for him at the observatory against his will.[35] Such quarrels indicate the problems in the context of changes that were initiated by the Academy during that period. Although one change was trying to find qualified employees, Christfried's assistants obviously could not replace the family members adequately. The working system of the household functioned inexpensively and smoothly, but was seen as possible competition for the Academy. The intention of the Academy was to establish the observatory as the only place where research was performed and to install the astronomer with his assistants there as its official representatives. Both functions should only be fulfilled by men.

These attitudes must also be understood in connection with the meaning of "honor" at the time. According to contemporary thinking, the Academy was seen as an institution, which was honored like an individual. To honor the institution also meant to honor the King. As with other institutions—the universities as well as the guilds—this honor was produced at the Academy in the form of representations. Women were excluded from the official representation.[36] On the other hand, the Academy could not afford to abandon the "household" astronomy. Thus, whereas the traditional observation practice had by no means been replaced by a new one, a clear division of labor—based on gender—had occurred.

## The Sister Household: Strategies and Arrangements

The death of Christfried in 1740 could have terminated the relationship between the Academy and the Winkelmann-Kirch household. The contact with the Academy, which had been established via the male head of household, should have been interrupted. The sources for this period, in reference to our questions, are incomplete and unorganized, but they show that this did not happen. In the Academy files, the sisters are mentioned as petitioners of the "quarter," an amount of money paid quarterly to widows and surviving dependents.[37] Extraordinarily, they received support from the Academy for a whole year when widows and surviving dependents of other scholars listed in the same files did not. They immediately had offered to carry on their brother's calendar production. This explains the relatively generous support by the Academy. In order not to put the Academy's finances at risk, the sisters had to be allowed to continue their brother's work. The temporary solution developed into a collaboration that lasted for decades. For a second time, a household led by women worked for the Academy. A few years after Christfried's death, the sisters had no official titles but received a regular salary from the "Académie Royale," as it was called by Frederick the Great after 1744. The bookkeeping of this period is particularly revealing in this context, as the receipts indicate that during the first two years two different salaries were paid. In addition to twenty-five Reichsthaler on a quarterly basis for each sister, Christine also received one hundred fifty Reichsthaler per year for the production of the *Schlesischer Kalendar* (Silesian Calendar), the most lucrative project for the Academy. Two central areas of the domestic astronomical production become apparent here as far as their payments are concerned. The first payment was probably received for tasks related to astronomy, such as observation. What other activities besides observation and calculations could the sisters have provided to be paid more than a "Mechanikus"? In order to explain the second payment, we need to look at the historical background. Only after Prussia had won the war against Silesia did the calendar market there fall under the monopoly of the Academy. Thus, although the Academy officially only allowed men to represent and work for it, Christine unofficially provided the most important (and at the same time most profitable) service.[38]

The sister household was therefore still active even though, as far as the sources indicate, official observations for the Academy were now carried out exclusively by male "astronomers," and only in the royal observatory. Both Kirch sisters were also busy with observation work that had nothing to do with the Academy. For example, they noted scientifically significant events at the time, such as the appearance of a comet.[39] The sources indicate that their notes were at least partially intended for publication. It cannot be determined, at this point, which audience they had in mind or whether these studies were in fact published. As in other periods, personal contacts, visits, and observation sessions with other members or associates of the Academy took place. Furthermore, there are references to correspondence with astronomers outside of Berlin, as well as references to the sisters' visits to the royal observatory.[40] The sisters' astronomical household functioned on the margins of an emerging (male-oriented) scientific society, for which it was still providing necessary services.

According to the vast number of receipts of the Academy, the payment methods underwent a change in 1759, leading to a restructuring of the household. From then on, only Christine Kirch received a salary (she was paid two different salaries, and the services mentioned above were still paid separately). Other evidence indicates that, during these years, Christine became the official head of household, although its exact structure cannot be deduced from the existing sources. However, it becomes obvious that, as the titular head of household, she had to provide for the rest of the family—"Myself and my family," as she wrote in a petition to the council of the Academy.[41] Her duties as head of household serve as an important argument for her petitions for an increase in salary. Christine thus fulfilled her duties as head of household by carrying out important parts of the calendar work for the Academy and perhaps other projects as well. For this reason, another person became part of the Kirch household. At Christine's request, the Academy gave her money for a clerk to carry out the administrative work under her supervision.[42]

From 1769 Christine received an extra payment—one hundred Reichsthaler per year—together with her other three payments. Unfortunately, there are not any further sources and this extraordinary situation can only be inferred from the files mentioned above. The Kirch household received a yearly income of four hundred Reichsthaler (of which fifty were paid to the clerk), an amount that compared favorably to the salaries of other Academy members.[43] One wonders why this sum, like other salaries, was not recorded as one single amount, but instead divided and paid out in separate payments. Was it thus possible to hide the fact that a woman, who was identified only by name, was paid more than some men? As a matter of fact, her brother, as the official astronomer, had only been paid sixty-two and a half Reichsthaler per quarter (249 per year).

This lack of information about Christine's work for the Academy, which lasted for decades, contrasts with a letter written in 1772 that attests to her significance. The Academy officially thanked Christine for her service. (She had grown old and found it more and more difficult to carry out her research.) The letter continued. The Academy politely asked her to teach Johann Elert Bode (later a famous astronomer) the skills of calendar making.

He was advised to follow her instructions and to function as her assistant.[44] From this year onwards, the knowledge and skills acquired in the context of the household were thus directly conveyed to a male astronomer, who used this knowledge in his position within the new institutional environment of the working system at the Academy. Whether the household included other women cannot be determined at this point. Based on current research, however, it can be stated that throughout the eighteenth century, the Academy of Sciences in Berlin needed the astronomical work of the sisters' household in order to guarantee its financial survival and reproduction.[45]

## Gender: The Invisible Economy of the Household Production of Science

The female members of the Winkelmann-Kirch family worked for the Academy of Sciences in Berlin more actively and for a longer period of time than had been previously assumed. Even after Maria Margaretha's petition to the Academy was rejected, and even after her death and the death of her son, Christfried, her daughters were paid for their services to the Academy. The Academy depended financially on the sale and distribution of the calendar, and the sisters were very much, if not solely, responsible for its production. The cooperation between household and Academy, which lasted for decades, shows that we cannot speak of an exclusion of women from the sciences during the eighteenth century after all. Ten years after Londa Schiebinger's research on Margaretha Winkelmann-Kirch, the story can now be amended and some aspects can be seen differently. Based on her research material, Schiebinger pointed out the consequences of two conflicting "social trends." According to her, the typically German craftsman traditions had promoted women's participation in the natural sciences, as women had access to the secrets and tools of a trade through the "apprenticeship-system."[46] However, the Academy's rejection of Maria Margaretha had meant a "turning point" in Berlin. According to Schiebinger, this turning point has to be seen in the context of the general exclusion of women from the intellectual culture at the time, i.e., the old and the newly founded institutions.[47]

On the basis of new research, this essay presents a more complex evolution. The suggested "phase-model" allows the recognition of the specific and differentiated forms of the division of labor between the Academy, on the one hand, and the household, on the other, during the eighteenth century. By examining science as a system of production, this dimension can be seen in the relationship between household, gender, and science as follows: We are dealing with two interdependent work systems, which are characterized by a single economy of complex, flexible, and reciprocal-hierarchical interrelations. The relationships between these systems were not permanent and could be regulated by gender. The development of household production sites in connection with the institutional demands of the Academy shows that the scientific field of astronomy, the household, and gender were related to each other functionally.

This economy of scientific work at the Academy was an invisible one: it actually required an economy of the household as well. In the case of the Winkelmann-Kirch's household, a number of factors have to be considered. Besides the social background, a pastor's daughter's family environment, one also has to take into account Maria Margaretha Winkelmann-Kirch's pietism, evidenced in her personal correspondence and an influence on her daily life and actions. An essential basis of the contributions of all Winkelmann-Kirch women can be seen in the particular early modern work and production culture, in which "nourishment," i.e., expenses and income, meant working together, and was not seen as solely a male responsibility. Terms and categories such as "public" and "private" no longer explain the particular participation of women in the sciences in early modern times. Sometimes they obscure the fact that in these centuries the household was not perceived as the opposite of public life. We have to keep in mind that the household itself was the center of different public forms. One must consider the widely shared sociability: students who lived in homes with their teachers; visitors to collections in private homes, which were created much earlier than the formal institutions; as well as participants in academies founded by private citizens and those in scientifically interested and active circles. Furthermore, one has to consider more closely the correspondence between households: as the center of scientific travel culture, as a resource for publication of old and new work.

Generally, the early modern household of the seventeenth and eighteenth century must be understood as the basic unit for life and work, including all classes and all parts of society. Its forms of cooperation, governed by factors such as gender, social class, and religion to name but a few, are likely to have influenced the "classic" contexts of scientific work: the cultures of royal courts or aristocratic patrons, the culture of salons, the workshops of craftsmen and makers of instruments, and also the churches and monasteries. Yet to be researched is the extent to which the culture and working systems of the household were connected with other scientifically important institutions. If we understand the old Greek term *"oikonomein"* as "to household" between such cultures, we gain many interesting starting points for the history of science. As the invisible economy of astronomy shows, the household played a significant role in the coming-into-being of this area of knowledge—and it was very much marked by the category of gender.

For the Academy of Sciences in Berlin, which had to battle problems of changing leadership styles and interests, the interrelation between the household and institutional work systems proved to be very useful. The Academy could not abandon the services rendered by women. At the same time, the Academy attempted to avoid, hide, or deny any collaboration with the female scientists of the Winkelmann-Kirch family. The fact that the Academy depended on the household and its culture meant that the Academy had access to knowledge, skills, rooms, instruments, working force, and wisdom that it could not yet provide itself. From the women's point of view, the collaboration with the Academy was useful as well. The assignments for the Academy not only provided them with what they needed to survive, but also allowed them to continue their own research. The connection between household and Academy was guaranteed by the head of the household, ideally a man. The fact that Margaretha as well as Christine Kirch acted as heads of the household for long periods of time can be explained by the notorious financial and organizational weakness of the Academy and the reality that it was not unusual for women to head their own households.

The long founding, transitional phase of the new institution in Berlin made it possible for a woman to receive payment from the Academy. These were forms of inclusion of women despite formal exclusion as members, forms of cooperation with women that maintained representational exclusion. The Winkelmann-Kirch women made use of their room for maneuver as long as they could. They responded to and adapted to the ever-changing demands of the growing—and therefore limited—academic institution. Mother and daughters used contemporary strategies: for example, Maria Margaretha acted as the needy widow. Nowadays, this might be misunderstood as pure subservience. The approach suggested in this essay also makes clear that, until Christine Kirch's death, an almost uninterrupted succession of women within the same family made the astronomical work possible. In the case of the Winkelmann-Kirch family, the organization of the household made a female scientific succession possible because, in this case, the education was transferred through women over two generations. At the same time, their history indicates which possibilities were made impossible as maleness became the deciding factor for participation in the scientific community. The women's working context could not be transferred and stabilized in the new institutions of modern science. Instead, the women of the Winkelmann-Kirch family were confronted with the ever-growing limitations placed upon them. For over a century they could not actively and openly participate in the Academy, a project that largely had been made possible by their work. Between the two production systems, gender functioned as the regulator of an eventually closed off institution that used female research, but officially only allowed men to participate.

Further research might reveal that the Winkelmann-Kirchs were an exception. However, this does not seem to be the case. Other research has shown that a number of women worked for the Academy of Sciences in Berlin.[48] The "astronomical household" was not an exclusively German institution. Future research could include comparison with other institutions in London, Paris, and Bologna, where women (and some men) were actively involved in astronomical research, and whose work can be analyzed in terms of the "household model."[49] Whether the concept of an "invisible economy" serves as a model only in the context of astronomy, or could be applied to other fields as well, has yet to be determined.

## Notes

1. Archiv der Berlin-Brandenburgischen Akademie der Wissenschaften (AAW Berlin), NL Kirch, Nr.6: *Beobachtungstagebuch Maria Margaretha Winkelmann von 1713,* f. 8.

2. While research has been done on the mother, her daughters are less well known and her stepdaughter, Theodora, is completely unknown.

3. The first scholar who dealt with Maria Margaretha Winkelmann-Kirch and who mentioned the subject of "household" was Londa Schiebinger in "Maria Winkelmann at the Berlin Academy: A Turning Point for Women in Science" *Isis* 78 (1987). Her research focused on this astronomer, whose rejection by the Academy after the death of her husband, Gottfried—an astronomer of the "Societät"—marked, according to Schiebinger, the "turning point for women in science," i.e., the exclusion of women from academic institutions. However, both mother and daughters continued to do their work for the Academy from 1720 to 1740, but now were identified as assistants to their son and brother, Christfried, who received the official recognition of the Academy. For references and support while searching for literature, I would like to thank Detlef Döring, Conrad Grau, Jürgen Hamel, Klaus-Dieter Herst, Klaus Klauß, and Londa Schiebinger.

4. See Beatrix Niemeyer, "Ausschluss oder Ausgrenzung? Frauen Im Umkreis der Universitäten Im 18. Jahrhundert," in *Geschichte der Mädchen- und Frauenbildung,* ed. Elke Kleinau and Claudia Opitz (New York, 1996); and Beate Ceranski, *"Und sie fürchtet sich vor niemandem": die Bologneser Physikerin Laura Bassi* (Frankfurt am Main, 1996).

5. The literature is vast. See, for example, Pnina G. Abir-Am and Dorinda Outram, eds., *Uneasy Careers and Intimate Lives: Women in Science, 1789–1979* (New Brunswick, 1989). Bennholdt Thomsen, Anke Guzzini, and Alfredo Guzzini, "Gelehrte Arbeit von Frauen. Möglichkeiten und Grenzen Im Deutschland des 18. Jahrhunderts," *Querelles* 1 (1986); Beate Ceranski, "Wissenschaftlerinnen in der Aufklärung: Überlegungen zu einem vergleichenden Ansatz," in *Geschlechterverhältnisse in Medizin, Naturwissenschaft und Technik,* ed. Christoph Meinel and Monika Renneberg (Stuttgart, 1996); Katharina Fietze, "Frauenbildung in der 'Querelle des Femmes'," in *Geschichte der Mädchen- und Frauenbildung,* ed. Elke Kleinau and Claudia Opitz (New York, 1996); Ingrid Guentherodt, "Urania Propitia (1650)- in zweyerlei Sprachen: Lateinisch- und deutschsprachiges Compendium der Mathematikerin und Astronomin Maria Cunitz," in *Res Publica Litteraria,* ed. Sebastian Neumeister and Conrad Wiedemann; Bärbel Kern and Horst Kern, *Madame Doctorin Schlözer: ein Frauenleben in den Widersprüchen der Aufklärung* (München, 1988); Lisbet Koerner, "Goethe's Botany: Lessons of a Feminine Science" *Isis* 84 (1993); Ingrid Kuczynski, "Reisende Frauen des 18. Jahrhunderts. A Nonconformist Race?" *Feministische Studien* 1 (1995); Barbara Shapiro, "Early Modern Intellectual Life: Humanism, Religion and Science in Seventeenth-Century England" *History of Science* 29 (1991); Ulrike Weckel, Claudia Opitz, Olivia Hochstrasser, and Brigitte Tolkemitt, eds., *Ordnung, Politik und Geselligkeit der Geschlechter Im 18. Jahrhundert,* Supplementa 6, *Das Achtzehnte Jahrhundert* (Göttingen, 1998); Heide Wunder, *Er ist Die Sonn', Sie ist der Mond: Frauen in der Frühen Neuzeit* (München, 1992).

6. On natural philosophy and the experimental household, see Lynette Hunter and Sarah Hutton, eds., *Women, Science and Medicine, 1500–1700: Mothers and Sisters of the Royal Society* (Stroud: Sutton, 1997); see also Alix Cooper, "Home and Household as Sites for Early Modern History of Science," in *Wonders and the Order of Nature, 1150–1750,* ed. Lorraine Daston and Katharine Park (New York, 1998), which introduces the household into the historiographic discussion of science as an essential sphere of scientific work.

7. On the history of the family and household production, see Michael Anderson, *Approaches to the History of the Western Family, 1500–1914* (New York, 1980); Katrina Honeyman and Jordan Goodman, "Women's Work, Gender Conflict, and Labour Markets in Europe, 1500–1900," *Economic History Review* 44, no. 4 (1991): 608–28; and Manfred Lemmer, "Haushalt und Familie aus der Sicht der Hausväterliteratur," in *Haushalt und Familie im Mittelalter und in der Frühern Neuzeit: Vorträge eines Interdisziplinären Symposiums 1990,* ed. Trude Ehlert (Sigmaringen, 1991). Michael Mitterauer, *Familie und Arbeitsteilung: Historisch vergleichende Studien* (Köln, 1992); Claudia Opitz, "Neue Wege der Sozialgeschichte? Ein kritischer Blick auf Otto Brunners Konzept des 'Ganzen Hauses'" *Geschichte und Gesellschaft* 20 (1995): 88–98.

8. On the household as a site of education, see Deborah Simonton, "Apprenticeship: Training and Gender in Eighteenth-Century England" in *Markets and Manufacture in Early Industrial Europe,* ed. Maxine Berg (New York, 1991).

9. Heide Wunder, "'Jede Arbeit ist ihres Lohnes wert.' Zur geschlechtsspezifischen Teilung und Bewertung von Arbeit in der Frühen Neuzeit," in *Geschlechterhierarchie und Arbeitsteilung: Zur Geschichte ungleicher Erwerbschancen von Männern und Frauen,* ed. Karin Hausen (Göttingen, 1993).

10. See Theresa Wobbe, ed. *Frauen in Akademie und Wissenschaft: Arbeitsorte und Forschungspraktiken, 1700–2000* (Berlin, 2002).

11. A considerable part of the family's unpublished work can be found in AUB Basel, Bestand L Ia 699 (Epistolae Kirchiorum) (Gothanus). See also Detlef Döring, *Der Briefwechsel zwischen Gottfried Kirch und Adam A. Kochanski, 1680–1694: Ein Beitrag zur Astronomiegeschichte in Leipzig und zu den deutsch-polnischen Wissenschaftsbeziehungen (Abhandlungen der Sächsischen Akademie der Wissenschaften zu Leipzig)* (Berlin, 1997), vol. 74, H. 5, 9ff.

12. Diedrich Wattenberg, "Zur Geschichte der Astronomie in Berlin im 16. bis 18. Jahrhundert I," *Die Sterne* 48 (1972), wrongly emphasized that Theodora did not participate in her father's work. See Döring, *Briefwechsel;* and University Library Leipzig, UB Leipzig, Bestand 01322, 196c.

13. See "Des Vignoles, Eloge de Madame Kirch À l'occasion de laquelle on parle de quelques autres femmes et des un paisan astronome," *Bibliotheque Germanique ou Histoire Littéraire de l'Allemagne, de la Suisse, et des Pays du Nord* 3 (1722).

14. UB Leipzig, Bestand 01322, f. 32r-44b/R contains Christoph Arnold's letters and notes addressed to Gottfried Kirch. Maria Margaretha Winkelmann, however, is not mentioned.

15. This separate area of study connected with astronomy has not been mentioned in research on Maria Margaretha Winkelmann-Kirch. See in this context, Döring, *Briefwechsel.* On the couple's meteorological work, see G. Hellmann, ed., *Das Älteste Berliner Wetter-Buch 1700–1701 von Gottfried Kirch und seiner Frau Maria Margaretha Geb. Winkelmann* (Berlin, 1893).

16. Döring, *Briefwechsel,* also concludes this and describes Gottfried Kirch's pietistic contacts at length.

17. According to the sources, she also appears as Maria-Margaretha.

18. Conrad Grau, the historian of the Academy, believes that there was another married sister. It could have been Theodora. On the other hand, Klaus-Dieter Herbst also mentions another daughter, whose name was Johanna. See Klaus-Dieter Herbst, "Gottfried Kirch (1639–1710)," *ASTRONOMIE + RAUMFAHRT* 37 (2000).

19. On the interest in astronomical, astrological, and similar studies and their integration in a religious context during the eighteenth century in Germany, see Rainer Baasner, *Das Lob der Sternenkunst: Astronomie der deutschen Aufklärung* (Göttingen, 1987).

20. Döring, *Briefwechsel.*

21. Martin Brecht, ed., *Geschichte des Pietismus: Der Pietismus vom siebzehnten bis zum frühen achtzehnten Jahrhundert,* vol. 1 ( Göttingen, 1993); Martin H. Jung, *Frauen des Pietismus: Zehn Porträts von Johanna Regina Bengel bis Erdmuthe Dorothea von Zinzendorf*

(Gütersloh, 1988); Martin H. Jung, "Mein Herz brannte richtig in der Liebe Jesu," *Autobiographien frommer Frauen aus Pietismus und Erweckungsbewegung: Eine Quellensammlung, Theologische Studien* (Aachen, 1999). Hans Schneider, *Der radikale Pietismus im 18. Jahrhundert* (Göttingen, 1995); Luise Schorn-Schütte, *Evangelische Geistlichkeit in der Frühneuzeit. Deren Anteil an der Entfaltung frühmoderner Staatlichkeit und Gesellschaft, dargestellt am Beispiel des Fürstentums Braunschweig-Wolfenbüttel, der Landgrafschaft Hessen-Kassel und der Stadt Braunschweig* (*Quellen und Forschungen zur Reformationsgeschichte* 62) (Gütersloh, 1996).

22. In Gottfried Kirch's observation diaries, Maria Margaretha Winkelmann-Kirch's results are quoted on a regular basis. Apart from various calendars, which carry her husband's name, she published three papers; see Schiebinger, "Maria Winkelmann."

23. These cannot be completely determined; they lived in Leipzig and in Guben (two hours from Leipzig), where the children were born.

24. Archive of the Franckesche Stiftung, Halle, D121 (52), f. 214–18 from 1692.

25. There was an "observatory" owned by the wealthy nobleman Baron von Krosigk (Korsigk) at Wallstraße 72. According to the observatory diaries, however, it was nothing more than a simple apartment.

26. Based on these diaries, which Kirch had to present at the end of each year, the working methods during this period can be reconstructed in detail. Since he had taken the position of astronomer, he not only recorded observations, calculations, and illustrations, but also reported—matter-of-factly—the cowork and assistance of other family members. Similarly, Maria Margaretha Winkelmann-Kirch's less extensive notes between 1713 and 1717 were reported, although they were obviously not produced for the Academy.

27. AAW Berlin, NL Kirch, Nr. 2, f. 54 (June 20, 1702).

28. Döring, *Briefwechsel,* 12ff.

29. Ibid.

30. Wolfgang R. Dick, "250 Jahre Berliner Sternwarte," *Die Sterne* 26 (1965).

31. The quotation at the beginning of this essay documents this cooperation; see AAW Berlin, NL Kirch, Nr. 6–8 (Maria Margaretha Winkelmann's observation diary from 1713 to 1717).

32. Jablonski, a secretary at the Academy, coined the title "observator." The title "astronomer" also did not correspond to any specific job definition. See a letter composed by Christfried, dated December 10, 1727, AAW Berlin, Bestand PAW, I-III-2, f. 43–f. 46R.

33. AAW Berlin, NL Kirch, Nr. 9-29 (Christfried Kirch's observation diaries from 1716 to 1740).

34. Other references, for example, of calculation methods which required two people, prove this assumption; AAW Berlin, NL Kirch, Bd. Nr. 26 from 1736 shows some of those observations.

35. For example, he complained about his assistant Georg Schütz, a mathematics teacher. Typically, Christfried argued that the older Schütz was not able to learn the necessary skills. This had to be acknowledged (if reluctantly) by the director of the mathematics classes; AAW Berlin, Bestand PAW, I-IV-37, for example, from 1717 and 1719.

36. Schiebinger in "Maria Winkelmann" describes Maria Winkelmann's expulsion in detail. The fact that the Academy saw its honor as explicitly gendered can be seen in the protocols taken at meetings. See, for example, AAW Berlin, Bestand PAW, I-IV-6, vol. 2, f. 214, dated April 9, 1716, and f. 269, dated August 18, 1717.

37. AAW Berlin, Bestand PAW, I-XVI-218 (receipts).

38. AAW Berlin, Bestand PAW, I-XVI-221 to 244 and 247 from 1749 to 1772 and 1776; see also Adolf Harnack, *Geschichte der Königlich Preußischen Akademie der Wissenschaften zu Berlin,* 3 vols. (1900; repr., New York, 1970), vol. 1, pt. 2, 489ff. The budget for 1776 documents the participation of "Mademoiselle Kirch [Christine]" at least until 1776.

39. AAW Berlin, NL Kirch, Nr.118 a, 118b.

40. On the international correspondence, see AUB Basel, Bestand L Ia 694 and Ia 697. The family's letters await further research.

41. AAW Berlin, Bestand PAW, I-XVI-218 (receipts), Fasz. 9 and 10.

42. Ibid.

43. AAW Berlin, Bestand PAW, I-XVI-221 to 244 and 247 from 1749 to 1776.

44. AAW Berlin, Bestand PAW, I-IV-5, f. 153.

45. According to Conrad Grau, one of Christine's nieces later married Johann Elert Bode, Christine's student and the Academy's astronomer and editor of the astronomical yearbooks. Thus, Grau does not agree with H. Clemens, "Die älterenen Ephemeridenausgaben der Berliner Akademie und die Begründung des Astronomischen Jahrbuchs," in *Festschrift für Prof. Dr. Wilhelm Foerster* (Berlin, 1902).

46. It must be noted that the semantic field of the English "apprentice" is wider than the semantic field of the German word *"Lehrling."*

47. Schiebinger, "Maria Winkelmann," 175ff.

48. During the eighteenth century, various women led projects such as the mulberry plantation, the basis for silk production and research. Though organized in an impressive way, it failed. In addition, female family members of other academicians can be found in the Academy files. Unfortunately, these files only reveal information by chance, and therefore provide only fragmentary details about the (working) lives of those women.

49. In the case of the famous German-British astronomers William and John Herschel, their sister and aunt, Caroline Herschel, continuously performed valuable work in their households. Because the Herschel men were both members of the Royal Society, Caroline Herschel was praised by the institution. This contemporary recognition explains her presence in modern histories of astronomy. Less attention has been given to Margaret Flamsteed, the wife of the first "Astronomer Royal" at the observatory in Greenwich; Maddalena and Teresa Manfredi, whose brother Eustachio, a leading Italian mathematician, brought them into contact with the Bologno Istituto delle Scienze (Institute of Sciences); the French mathematicians and astronomers Nicole-Reine Lépaute (1723–1788) and Hortense Lépaute, brought into contact with the Académie Royale in Paris by the astronomer Jerome Lalande; and Marie Jeanne Lalande, who married Jerome Lalande's nephew, Christian, and worked with him on projects initiated by the Académie.

GRIGORY A. TISHKIN

Translated by Albina Krymskaya

# Princess Ekaterina Romanovna Dashkova and Women's Issues in Russia in the Eighteenth and Nineteenth Centuries

Scholarly research on various aspects of Princess Ekaterina Romanovna Dashkova's (1743–1810) activities regularly appears in academia. This special interest in her life can be explained by several reasons: the celebration of the 250th anniversary of her birth, her achievements in supervising two scholarly establishments—St. Petersburg Academy and the Russian Academy of Sciences—and active interest in the Russian Enlightenment and Catherine the Great's times.[1] She was a truly remarkable woman.

Dashkova was descended from the Vorontsovs, an old, aristocratic family of imperial Russia. Many of the family members occupied high positions in the Tsarist government. Princess Dashkova was by Catherine's side in 1762 during the "Palace Crisis" (when Catherine seized power from her husband Peter III) and rendered significant moral support to the future empress. Quite naturally, the young Princess Dashkova hoped for future reciprocal support on the part of Catherine II. A poet who was her contemporary wrote that Dashkova's ambitions were most striking and not typical of the women of that time, and that she even embraced the desire of "becoming a member of the [proposed] Council of State."[2] At times, Dashkova went so far as to advise the Empress on personal matters. The Council of State was never called, and as a consequence of her boldness, Dashkova found herself banished from the Winter Palace for most of the 1770s. These were the long years of disappointments, travels throughout Europe, and the education of her son and daughter.[3] The relations between Catherine II and Dashkova began to improve only near the end of the decade. In

January 1783 the princess was appointed director of the St. Petersburg Academy of Sciences (founded by Peter the Great in 1724). Several years later Dashkova headed yet a second educational institution, the Russian Academy (founded by Catherine in 1783).[4] Through these two appointments she was able to play a significant role in the development of science and education in Russia until Catherine's death in 1796.

While much has been written about her time as "Madame Director" of the Academy, the issue of Dashkova's influence on the dissemination of ideas about women's equality in Russian society has not been researched enough. In this article then, we will try to explain why one of Russia's nineteenth-century scholars called the princess a "partisan of women's equality."[5] This explanation requires that we describe Dashkova's influence against a background of the public interest in the women's issue in Russia in the second part of the eighteenth century.

## The "Women's Issue" and Women's Education

Beginning in the middle of the eighteenth century the idea of improving women's status in the family and society found its supporters both in the capital and gradually in the provinces. Contemporary women's attitudes were influenced by new images of women keen on scholarly and artistic activities, such as authors, poets, and public women not only from eighteenth-century Russia and other European countries but also from former times.

In the 1770s in Russia there appeared a publisher with great courage who took liberties to serve the interests of women readers with a special women's magazine entitled *Modnoe ezhemesiachnoe izdanie, ili Biblioteka dlia damskogo tualeta* (Fashion Monthly Issue, or Library of Ladies Dressing) (1779). This periodical for women and about women was far removed from tearful sentimentality and hypocritical moral tales. Women's rights to liberal feelings were popularized by the magazine. The publisher told women readers about their public significance and even about their possible superiority to men. Posing those questions and answering them, N. I. Novikov, the publisher, wrote the following:

> Who is more useful, a man or a woman?
> Hero strives to protect his Fatherland,
> Councilor of State to spread the truth in court
> Of woman born a hero and a councilor
> Who is of more use please you decide.[6]

N. I. Novikov's magazine was a significant event in women's journalistic history because its goals were to educate, to train, to entertain, and to inform Russian women readers. N. I. Novikov is well known in the history of Russian public thought not only as the most prominent disseminator of enlightened ideas but also for his great services as a founder of one of the first women's magazines. This edition had several tens of subscribers. Later on, in 1791, it was replaced with another magazine entitled *Magazine of English, French and German Fashion with attached descriptions of the way of life, public entertainments and pastimes at the most noble cities of Europe, with welcome funny stories,* which began to be published in the same Moscow University printing house. This magazine had existed about a year. The distinguishing feature of this edition in comparison with Novikov's first creation lies in the fact that it was not only for women, though there were more articles for women. Since 1798 one more magazine had appeared: *Zerkalo sveta* (Mirror of Society). It is interesting to note that upon first acquaintance with these periodicals, today's readers would be alternately embarrassed and surprised by the sometimes flattering and sometimes arrogant way of talking to women. However, as it seems to us, this language was appropriate to the spirit of literature and the journalistic language of that epoch. In the second part of the eighteenth century patriarchal severity existed alongside gross dissoluteness.

New, easy language that was more understandable and appealing to women readers was introduced into literature and public writing by N. M. Karamzin. The language in which his works were written attracted many women readers, who in their turn made Karamzin's name and works popular and therefore widely read. Beginning in 1791, N. M. Karamzin, a prominent Russian historian and publisher, in *The Moscow Magazine* often popularized the ideas of Rousseau's *Nouvelle Heloïse.*[7] He attempted to interpret women's problems in different genres using history and belles lettres, for example, in the short work *Historical Eulogy to Empress Catherine II* (1801) and in the heroic image of the real historical person in *Marpha Posadnitsa, or Subdual of Novgorod* (1803). Young noble girls not only had read about but also had dreamt of loving passion with Richardson's *Clarissa,* and with Karamzin's *Aglaya* and *Poor Liza.* A generation of young women readers devoured not only the easy, sentimental, and moralizing belles lettres but also popular scientific works. Noblewomen became active in public, educated themselves, and were inspired to believe in their strength and abilities.

Books and magazines suggested the ideal women's image to female readers on country estates and in private residences in towns. In the first issue of *Vestnik Evropy* (Europe's Herald), Karamzin gave a general portrait of the contemporary woman: "an ideal woman does not have to be beautiful, the most important is that she has to be kind and nice, which can be achieved only by reading. This opinion is not exceptional but very widespread."[8] Following Novikov and Karamzin, the same opinion was popularized by P. I. Makarov, publisher of *Moscow Mercury.* In the preface of his journal he not only set the task of pleasing women readers but also asked the question: "Why would a woman become as educated as a man? Terrible are those people who think that women lose the beauty of their sex acquiring knowledge."[9]

Many periodicals at the end of the eighteenth and beginning of the nineteenth century followed the road that was opened by Novikov's magazine for women, *Modnoe ezhemesiachnoe izdanie, ili Biblioteka dlia damskogo tualeta.* Among them were: *Magazin angliiskikh, frantsuzskikh i nemetskikh novykh mod* (Magazine of New English, French, and German Fashion) (1791), *Moskovskii zhurnal* (Moscow Journal) (1791), *Vestnik Evropy* (Europe's Herald) (1802), *Zhurnal dlia milykh* (Journal for Pretty Ladies) (1804), *Drug prosveshcheniia* (Enlightenment's Friend) (1804), *Moskovskii zritel'* (Moscow Spectator) (1806), *Liubitel' slovesnosti* (Lover of Philology) (1806), *Aglaia* (1808), *Drug iunoshestva* (Youth's Friend) (1808), *Ulei* (Beehive) (1811), *Ukrainskii vestnik* (Ukraine's Herald) (1816), *Modnyi vestnik* (Fashion's Herald) (1816), *Damskii zhurnal* (Ladies' Journal) (1823), and others. They actively joined in discussion of the women's issue. Despite having different goals and social audiences, first of all, each attempted to widen the mental outlook of female readers, who listened with passionate curiosity, read carefully, and watched what was going on in the world. Secondly, those journals not only wrote for women and about them but also urged Russian-educated women to participate in creative activity. They attempted to motivate women to write, to invent stories, and to create.

Many public figures and memoir authors of the second half of the eighteenth century stressed the importance of women's education. To prove this, let us describe several examples. In his *A Word on Education,* presented at the ceremonial reception of scholars, professor of Moscow University A. A. Prokopovich-Antonsky not only maintained that "the fortune of nations depends very much on the education of young people," but he put the issue of women's education among the most important requirements for success. Prokopovich-Antonsky insisted that "society's benefit" required "change in the physical and moral education of women." "Why," asked the orator of those present, "did women have such a weak build, such pale faces, such a great number of diseases? Why, if not because of the kind of education they received?" Women, in professor Prokopovich-Antonsky's words, often complain about boredom, "killing their time living in idleness and inaction, at meetings of idle talk and gossip." The reason for this, in the professor's opinion, was lack of "useful information."[10]

A. A. Prokopovich-Antonsky was not the first to question women's role in Russian society. Twenty years before that, his colleague, a well-known professor of the same university, S. E. Desnitsky, at a crowded meeting of professors gave a speech stating that the progress of human society required equalization of "the female to the male sex, and in some cases giving preference to the female." Like other contemporaries, Desnitsky linked women's emancipation to the spread of education, culture, and knowledge: "Only by educating people will the government favour equalization and praise of this sex overthrown in ancient times."[11] In the last quarter of the eighteenth century tendencies similar to those at Moscow University could be seen at St. Petersburg University. In their joint monograph on the history of St. Petersburg University, supported by M. De-Pule's memoirs and protocols of St. Petersburg Academy's Conferences, Iu. D. Margolis and

G. A. Tishkin give examples of women's attendance at professors' lectures.[12]

Noblemen, outside the universities, also spoke on behalf of women. In the second half of the eighteenth century, in connection with the beginning of industrial development and the dissemination of Enlightenment ideas, and with the aggravation of ideological battles and the fact of peasants' protest against the reinforcement of feudal customs, the progressive-minded representatives of Russian society involved in the creation of the nation and a national culture spoke in support of women's equality in education and participation in public life. However, the movement for women's emancipation in those times had its ideological restrictions. Only a gentlewoman was considered a "woman." Women of other social groups were out of sight even for progressive thinkers of that epoch.

What was the reason for the expression of these ideas? Was it because of the authors' feelings of sincere and motiveless admiration for women? The authors probably pursued other goals. The most important for most of them was popularization of the idea of the nobility's superiority over other groups in Russian society. Under other circumstances, authors eulogized a woman in their poetical fantasies but wrote abusive characterizations of women in letters to close friends and in their memoirs. However, it is important in this connection to note that public morality in the second part of the eighteenth century changed so much that for an "enlightened" person or for someone trying to confirm that he was enlightened it was very dangerous to abuse a woman in public. Public opinion—read as "women's opinion"—decided who was and who was not enlightened.

E. Shchepkina, the well-known author of the history of women in Russia and professor at Bestuzhev Courses (Women's Higher School), was right when she noticed the particular conflict for women of this era: the privileged status of a few notable women in a society that overall still enforced women's subordination to the stronger sex and to prescribed customs.[13] Thus, whoever spoke on the women's issue in the second half of the eighteenth century had to be courageous and committed. It was hard to hear the voices of women; sometimes they took the more traditional patriarchal position; they were afraid of innovations. Decades would pass before some women would find the strength and courage to announce their intentions and to uphold them by establishing societies for the advancement of women.

## The Role of Princess E. R. Dashkova

Of the male authors who supported resolution of the women's issue, we think those deserving of detailed attention are those whose activity left outstanding tracks in the history of Russian public thought. In our opinion those are I. I. Betskoy, S. E. Desnitsky, and P. A. Alekseev. But E. R. Dashkova played the most important role by publishing some authors' works. In 1871 the famous *Russian Archive* published the collection of letters addressed to Ivan Ivanovich Pamfilov, private confessor to Catherine II. Thanks to regular and continuous talks with his ecclesiastical daughter—the Russian Empress Catherine the Great—this man held an exceptionally important place in the church hierarchy of the Russian Empire. He sought to maintain friendly relations with all the leaders of the Russian Orthodox Church and even with statesmen such as G. A. Potemkin, A. A. Bezborodko, and many others. His influence was sought by notables all over Russia, because personal contact with I. I. Pamfilov meant a lot in decisions about complicated official and private cases. Among those letters published in *Russian Archive* were several from an archpriest from Moscow's Archangel Cathedral (situated in the Kremlin). But what relation do Catherine's confessor and the archpriest from Moscow have to the women's issue? A very direct one, as we will see further.

The dean of this famous Kremlin cathedral appealed to the Empress's confessor on the issue that is the theme of our work. He wrote:

> All-honest father archpriest Ioann Ioannovich, Sir! There is a foreign saying that: "who is absent is to blame." So being 70 verst away [1 verst = 3,500 feet], I am blamed for the publication in St. Petersburg of the translation: *On the Nobility and Advantage of the Female Sex*. It was found in letters after my brother died and last year, for curiosity's sake, sent by me to E. R. D. [Ekaterina Romanovna Dashkova], not only without my approval of its contents but also with my censure of the absurdities of the original. Hence I did not even imagine that somebody could publish such a work or that I would receive the undeserved reprimand from the Empress's retinue. God saved me from the consequences of this disapproval, thanks to your representation which I heard of via a letter of your fellow member S. M. Kor and via information from Mr. V. I. Bazhenov, newly arrived here in Moscow, for which I thank you sincerely, and while there is a spirit of life in me, I will not stop thanking [you] such a gracious protector.[14]

This letter is rich in content, and with a full deciphering of hints and references it puts a reader into the thick of events of the last third of the eighteenth century, into the relationships between people who spoke very actively on the women's issue. First of all, we will pay attention to "E. R. D."—Princess Dashkova—on whom the author of the letter to I. I. Pamfilov shifted the responsibility for the appearance of the book *On the Nobility and Advantage of the Female Sex*.

In 1783 Catherine II made E. R. Dashkova director of St. Petersburg Academy. In October of the same year the Empress's decree established the Russian Academy, and Princess Ekaterina Romanovna (or "Madam Director" as she was called by the staff of St. Petersburg Academy and the University students) was appointed as its president as well. To be director meant supervising of what was an educational-research institute consisting of three interconnected parts: a gymnasium for tuition in languages; a university for philosophical and scientific instruction (to which students could transfer after finishing their linguistic studies); and an academy (or institution of scientific research, which would provide employment for university graduates). Dashkova's long and successful service at this state post proved women's ability to work effectively for the commonweal, whereas before her men could not cope with the requirements of the job.

Dashkova's pedagogical experience began in Scotland, in Edinburgh University, where her son studied in 1776. As the son studied, so did his mother. A young, thirty-year-old woman, full of strength, with an inexhaustible thirst for knowledge, was watching, reading, and regularly communicating with the best scholars of Great Britain. She could not foresee that very soon after coming back to Russia she would be the head of the St. Petersburg Academy and Russian Academy, and that her knowledge of the organization of higher education and scientific researches in England, Scotland, France, Italy, and the German states would be useful for improving university education in St. Petersburg.

When she took over in 1783, while mindful of the achievements of former directors, she set about introducing what she saw as the best aspects of European universities into the Russian educational system. She expelled students "on the grounds of their absolute incapacity," as she explained later in her *Memories*. All told, Dashkova wrote in her *Memories*, "I found only seventeen pupils [gymnasium-level students] and twenty-one students of arts [university-level students]; in my time the number of the former was increased to fifty, the number of the latter to forty."[15] In her report to Catherine II in 1786, she made the same observation, not only claiming that she had increased the number of pupils to eighty-nine, but that the students were now "fed, dressed, and educated much better than before."[16] There is no doubt as to the reliability of these data, and the success is quite impressive, considering the low starting point.

The staff was also radically changed. The musician who taught the violin at a salary of eight hundred rubles a year was dismissed, and the money was used to employ a tutor and teachers of English and Italian.[17] English and Italian were new departures. Hitherto the staple diet of the gymnasium-level curriculum had been Latin and Greek, supplemented by German and French. The "Proceedings" of the Conference of the Imperial Academy of Sciences bear witness to her special interest in languages and translation: she fired teachers she deemed uncommitted. According to a minute of March 1784, there were to be eight students "capable of perfecting their knowledge [of languages]," and "Madam Director" required that the co-Rector, Hackmann, see to it that they study logic, Latin, and Greek. It appears, though, that Hackmann fell short of "Madam Director's" expectations and was not conscientious enough in conducting classes in the two classical languages. As a result, he was dismissed, so that "the students should not be without proper teaching any longer."[18]

Dashkova taught some students herself. That is why alongside developing the curriculum, establishing strict examination requirements, and enforcing discipline among students and staff, she paid much attention to the work of special groups of students who translated books into Russian. To foster this work, she created the Assembly of Translators (later the Translation Department). Consisting mostly of university students, it provided a good opportunity for practical language study. Language teaching was Dashkova's special concern. The development of science in the eighteenth century required this, and today Dashkova's ideas still hold. We know from the moment she took over the Academy, Dashkova herself taught languages to some of the students. D. A. Tolstoy states that on 30 January 1783 Dashkova introduced a rule by which each student, in turn, was to spend the whole day (from 8 a.m. to 7 p.m.) with her, this practice allowing her to know the abilities and character of each pupil. In the morning Dashkova taught students to write and translate. Unfortunately, we do not know how long this innovation lasted.[19] She also made it possible for the best students at the gymnasium to study abroad, at Göttingen University in Germany. In addition, as head of the Academy of Sciences, Dashkova not only supervised the development of new forms of examination but also attended the examinations herself and would listen attentively to the candidates' answers.

One more of Dashkova's pedagogical undertakings as director of the Academy was resumption, with financial support, of the wonderful tradition of public lectures in the Russian rather than the Latin language, which had been introduced when M. V. Lomonosov supervised St. Petersburg University. This worthwhile practice was kept even after Paul, Catherine II's successor, exiled Dashkova as he did others among her courtiers. The first discussion of public lectures took place at the conference of the Academy of Sciences with Dashkova's participation on 3 July 1783, documented in the "Protocols." The author of the article on Dashkova in the *New Encyclopedic Dictionary* noted that the idea of public lectures was suggested in her speech given upon entering her duties in January 1783. He referred to Dashkova's words indicating that science would not be the monopoly of the Academy but would "belong to all the Fatherland and taking roots would prosper." Also to further Russia's scientific progress, Dashkova recommended membership in the Academy for foreign scholars, such as the chemist and physician Joseph Black and the historian and rector of Edinburgh University, William Robertson. Members of the Academy and adjuncts voted for them unanimously.[20] Robertson's work was definitely highly thought of in academic circles. In the 1770s and 1780s his historical studies were translated by students of St. Petersburg University from English and French into Russian and appeared in the periodicals and other works published by the Academy.[21]

Everyone who has written of Dashkova has praised her comprehensive abilities. However, those many researches on the princess have not considered her thinking on the women's issue, even though during the life and state service of "Madam Director" ideas of emancipation and of the equality of women in public affairs and in the family were widespread in Russia, especially in St. Petersburg and Moscow.

In this essay, we suggest that one of those who paid attention to Dashkova's special thinking about women's emancipation was the author of a scholarly, popular biography of the princess, published after the appearance of the selected letters to I. I. Pamfilov, the Empress's confessor, in *Russian Archive.* The nineteenth-century author V. V. Ogarkov dedicated all his work to Dashkova's well-known talents for ruling both academies; he described her as a "passionate partisan of women's equality."[22] The author used the word "partisan" in the sense of "supporter." So Ogarkov's words should be read as "E. R. Dashkova was a passionate supporter of women's equality." We doubt that Ogarkov came to this conclusion by only researching documents on the women's issue in Russian public thought in the second half of the eighteenth century. Besides P. A. Alekseev's letter thanking Pamfilov in *Russian Archive,* Ogarkov could find this information on Dashkova and the women's issue in M. I. Sukhomlinov's and N. I. Barsov's works.[23] Among contemporaries of the second half of the eighteenth century, archpriest P. A. Alekseev had a reputation as a liberal and lover of freedom.

As for his brother, I. A. Alekseev (1735–1780), he also served as a priest in one of Moscow's churches but then spent his last years in St. Petersburg lecturing on law to the Polish Gentry Corps. Both brothers did translations and were not strangers to scholarly research. But the answer to the question of whether the deceased brother of a Moscow archpriest translated a book by the German Humanist scholar Agrippa Nettesheim (1486–1535), *On the Nobility and Advantage of the Female Sex,* is still unclear. What interests could he have had? And how could papers of the late law instructor of the Polish Gentry Corps travel to Moscow and later from there to St. Petersburg?

Previously, neither the law instructor, I. A. Alekseev, nor his brother, the archpriest of Archangel Cathedral, P. A. Alekseev, were known to have taken an interest in this subject or in the history of the women's issue. Today, well-known scholar and philosopher Iu. N. Solonin has given his attention to this problem. His in-depth article "Voice from the Eighteenth Century in Favour of Women's Dignity" is dedicated to the analysis of the translation of Nettesheim, as well as to establishing its authorship.[24] We support Solonin's doubt that the author of the translation was the deceased brother of the Moscow archpriest. It is more likely that P. A. Alekseev, after getting into trouble because of the "absurdity" of the translated text and after receiving what he believed to be an "undeserved reprimand" from the Empress's intimate courtiers, wanted to exonerate himself. In such cases in the eighteenth century, as well as in the twentieth century, usually one could lay blame on the dead even though on the book's title page was printed: "this book was translated in Moscow under the guidance of the archpriest of the Moscow Archangel Cathedral Petr Alekseev." In this case it turns out that the Moscow archpriest must have lied at least once: either when his name was published with the book title or when he wrote to the Empress's confessor that the translation was "found in letters after my brother died and last year for, curiosity's sake, sent by me to E. R. D." Iu. N. Solonin in his article does not comment on why the translation was handed specifically to E. R. Dashkova, on why the Empress's confessor, I. I. Pamfilov, protected P. A. Alekseev, or on why V. I. Bazhenov was the one who brought out the details of the whole scandal after the publication of *On the Nobility and Advantage of the Female Sex*.[25]

Looking back to the history of the establishment of the Russian Academy and Dashkova's work then, the situation can be explained. It is worth becoming acquainted with the list of Russian Academy members in 1783–1841, published as Appendix 2 to the monograph by V. V. Kolominov and M. Sh. Fainshtein.[26] We discover that during the first two years of existence of this famous academic institution headed by Dashkova, the Russian Academy had fifty-four members. Those who interest us are I. I. Pamfilov and P. A. Alekseev, who rank in the list one after another (election year—1783), and V. I. Bazhenov, who was among those who became members in 1784. Thus, all of them had known each other before the publication of *On the Nobility and Advantage of the Female Sex*. As the natural consequence, the manuscript of the translation came into Dashkova's hands and was published in the printing house of the Academy of Sciences—of course, not without her approval. So V. V. Ogarkov was right when he characterized E. R. Dashkova as a "passionate partisan of women's equality."

## The Treatise and Its Translator

The structure of Alekseev's translation is very simple and clear: it consists of eight theses on "the advantage of the female sex" and some arguments to support them. Today those arguments would lack any special strength because all of them are taken from Holy Writ, classical authors, and history. But to Russian readers of the eighteenth century this argumentation was natural; such proofs were clear to them, and they acted on people's opinions of that time like axioms. One should remember that the translator (or translators), headed by the priest, attached special importance to Holy Writ despite some secularization of Russian society, a process that would not be completed during the eighteenth century. In addition, church ideology permeated even those areas of intellectual activity that were far from church life—law, philosophy, art, journalism, and literature. Others used church rhetoric widely in all areas of life.

Did church ideology negate in contemporaries' minds advocacy of the improvement of the social and family status of women? No, Christian dogma could also be decisive for support of the women's issue both in the second part of the eighteenth century and into the nineteenth century. It made contemporaries understand more easily and assisted in the penetration of those dogmas into people's hearts and minds, especially women's. Both in the middle of the eighteenth century and a century later, during the years of the first revolutionary situation in 1861, church representatives, its press, and journalists took an active part in discussion of the women's issue, admitted the necessity of changing women's status, and based their views on biblical principles. For example, in 1862 the priest D. Sokolov considered women's interest in progressive ideas and their aspiration for equality as nothing more than "pleasing to false taste." Even so, here is how he explained the Christian concept of relations between the sexes in his work on women's purpose: she (a woman) is not lower than the man because she is not only a husband's help but she is a help "similar to him," for only in a condition of equality could she give him the help he needs. However this place is secondary and dependent: a wife was created after a husband—was created for him. As she was taken of him she is "bone of his bones, flesh of his flesh," and "she is joined with him so closely that he can not abase her without abasing himself, but at the same time being created from his rib she owes him her world which she enjoys and from which she has her name."[27]

Agrippa's work as presented by P. A. Alekseev begins with confirming a thesis on perfect sexual equality. The author gives attention to physiological differences "which child-bearing required." As for the spiritual and mental aspects, the author does not see any difference and cannot find any advantage of one sex over another: "The same as her husband a wife has sense, mind, speech, the same bliss she seeks." However, "as for other things except the essence of the soul then the female sex excels over males endlessly."[28] If we compare this statement with the one that was cited above from priest D. Sokolov's work, although it appeared almost a hundred years later, the present reader can see that archpriest Alekseev's translation is unusual even for readers at the end of the twentieth century. And in 1862 when D. Sokolov's work appeared, Alekseev's opus *On the Nobility and Advantage of the Female Sex* had not been forgiven. The church press described it as a "bold book on the women's issue" that would not be permitted even in "other free countries," and reported that it had exasperated Catherine II.[29] One can conjecture that Alekseev's work exasperated the official church more than Catherine.

Thus, the treatise's author thought that not a human being in general but a human being in woman's image was the crown of creation. Alekseev used the scheme of creation in the study of the fourth-century church father St. Grigory Nyssa for confirmation of the female's "nobility and advantage."[30] In the process of the concrete definition of humanity, Agrippa/Alekseev divided the creation of human beings into two phases: the first phase was creation of the male, and the second of the female. Thus, he stressed her perfection and advantage: "and a wife is the last of creations . . . who could argue that she is not the most superior of all creatures." Here Iu. N. Solonin, the twentieth-century expert on this translation, comments on events described in the correspondence between P. A. Alekseev and I. I. Pamfilov from the *Russian Archive:* "so far we don't know the reaction of scholarly circles of the Russian church to this work and it is unknown if there was one to the text's idea, but secular circles, as we saw, responded quite severely, but motivating them there was their ignorance."[31]

In contrast to Solonin, we do not know of any reaction in secular circles. We are not going to argue about "their ignorance," but where is their reaction? Everything said in the article by the famous scholar of St. Petersburg University (Iu. N. Solonin) we see as the reaction of the church circles. It is easy to see their opinion in the correspondence that was published by *Russian Archive.* There are I. I. Pamfilov's references to the displeasure of the Empress and of some of her intimate advisors, but nothing about what displeased the Court. It is known that when Catherine II was displeased by Princess Dashkova, either because of the behavior of Dashkova's daughter or some inappropriate works published in the Academy Press, she expressed her displeasure to Dashkova in person. True, there were some situations when the Empress laughed angrily behind the back of the director of St. Petersburg Academy and caustically rebuked her for her children, but we failed to find anything similar about the treatise *On the Nobility and Advantage of the Female Sex.* Concerning Pamfilov, the Empress's confessor, one can suppose that he, as a member of the Church, could feign his own displeasure by presenting it as an opinion of the Court or the Empress's retinue. It is important to note that in reality Dashkova had many evil-wishers among the courtiers who were ready to use any occasion, any false step, to discredit her in the eyes of the Court and the Empress.

The treatise continues on the same theme, using arguments of church father St. Gregory of Nyssa on the superiority of the female. Eve was born in paradise while Adam was not, and was created from his rib, not from the earth. Such arguments make us smile, but the treatise's author did his best to see signs of women's superiority in everything. It is possible that it is here that critics interpreted his statements as "absurdities" and "curiosities."

Agrippa/Alekseev presents arguments and makes statements about women's "bodily beauty" that would have been blasphemous and sacrilegious. The author moves to examples from treatises by Pliny the Elder, Galen, and Avicenna. He contrasts women's scent, her cleanliness, and the lack of hair on her face with the unclean, ugly, beastlike man. He states that a woman can conceive a child without copulating. Thus, the miracle of the Immaculate Conception, an exceptional and unique sacred event, is introduced as an ordinary event, "known by Turks and Mohammedans."[32] In ten wordy pages, Agrippa/Alekseev argues that Adam, not Eve, was the initiator of original sin: "a husband drew on death, but not a wife, and all of us sin in Adam but not in Eve."[33] If the woman is the source of all blessings disseminated by God to mankind, then a man is the source of harm. For a male is the source of all unjust deeds, aggression, and death. Adam was the first who broke the Lord's commandment; Cain was the first man killer and fratricide; Lamech, the first bigamist, etc., right up to the men who were the first "who allied themselves with demons and invented nasty skills."[34]

The treatise's author discusses women's altruism, which is not to be compared with men's. As for their chastity and moral purity, then, "of women, excluding Virsavia only, we do not find anything else except that each woman was always satisfied only by her husband . . . for wives are much more temperate than husbands." This comment by P. Alekseev on the holiness of marriage and the faithfulness of wives to husbands sounds as if he lived in another, unreal world, knowing nothing about what was going on in Russia at the Imperial Court and as if, as confessor, he did not receive confessions from his parishioners. Was M. M. Shcherbatov the only one who knew about extramarital relations in Russian society, described in his work *On the Damage to Morality in Russia*?[35] "But who, I say, of males however old, cold, barren and unable he was would treat his wife so devotedly to let another male cohabit with her and that another male would fecundate her maw with his sperm."[36] In other words, only a woman decides to take others to fulfill her matrimonial duties and reproductive functions; men would never allow it.

The translation of Nettesheim went on to insist that woman was the best educator for the human race; she was possessed of exceptional abilities for sciences, crafts, and arts because of her more perfect nature, higher intellect, and perfection of feelings. She was the only one who was the beginning of morality, kindness, and beauty. In contrast, "husbands everywhere were rash, voluptuous, bigamists, polygamous, polyfalse, fornicators and lechers."[37] A woman, thought the treatise's author, was more capable of feeling justice and even had more ability for law than a man. In this part, Agrippa/Alekseev certainly meant no one but Catherine the Great: "could you imagine if women were permitted to create laws, write histories, what tragedies about the excessive villainous acts of men they would create!"[38] Further, Agrippa/Alekseev jumped more than two centuries into the eighteenth century and spoke of the most important issue: whether women could become priests. He thinks that it is not forbidden for her and that there were examples from Holy Writ. He explains the lack of women's status in the church hierarchy by its debarring her from education. And then Agrippa/Alekseev, like D. I. Pisarev later in the 1860s, claims that if women were allowed to study they would excel men in knowledge because women by their nature excel as inventors of science.[39]

Both here and in other passages, reading descriptions of the "female sex," one can catch oneself thinking that all of this could be included, as writes Iu. N. Solonin, in the most radical programs of the present time. The same was noted by one nineteenth-century author in his review, when he called *On the Nobility and Advantage of the Female Sex* a book containing a glorification of women that would be difficult to find in that time in our (Russian) literature or in foreign ones.[40] Such a response to Agrippa/Alekseev's book appeared one hundred years later, that is, in the mid-1870s, when Russian and foreign philosophy and public writings were filled with works on women's emancipation.

In the last argument of *On the Nobility and Advantage of the Female Sex,* the author points out that long ago ancient legislation protected women's dignity and honour, gave women and men equal rights, and sometimes advantaged women. But "males tyrannically against God's right and natural laws took over, then women's God-given freedom is lost to unjust laws, destroyed by habits and custom, and killed by education." Therefore, woman was created free by God and nature, but she lost her freedom owing to unjust social relations, in which violence and oppression ruled. "And in this way by the force of these laws women, as if defeated by war, must yield to victors not by natural or supernatural need or reason but by habit, fortune or some forced occurrence."[41]

Upon acquaintance with this treatise one finds oneself in the center of polemics in the *Sovremennik* [Contemporary] *Magazine's* pages from the middle of the nineteenth century and in works by the well-known fighter for women's rights—or the "lawyer of women's rights," as he was called by his female contemporaries—M. L. Mikhailov. He was sure that men's brute force alone became the cause of the violation of harmonious public relations and of women's enslavement. This idea runs through many writings by Mikhailov, even those that he wrote in exile in Siberia. For instance, working on a popular scientific essay on the life of primitive peoples entitled *Outside of History,* Mikhailov again and again returned to asking why one sex was oppressed by the other. He supposed that the dependent status of women originated long ago at the initial stages of mankind. "People were like animals. They stopped to be animals when a male began to value beauty in the female. But the female was weaker than the male in strength, and the male was crude. Everything was decided by force. The male appropriated for himself a female beauty which he began to value. She became his property, his thing."[42] In his novel *Together,* Mikhailov analyzed the problems of women's public and family dependence and the main point of love. He wrote the following: "Very deep inside of our light feelings there are tears. . . . Who has not counted in those tears, even if instinctively, the sorrow and memory about secular injustice that never loved. Who has not seen in women's eyes looking at him with trustfulness, the tears for life's genius suffered and fettered by secular chains, has never known what women's love was."[43] Perhaps the author of the eighteenth-century translation should be put in the same category with M. L. Mikhailov, the poet and publicist of *Sovremennik* and women's rights advocate of one hundred years later.

It seems obvious to us that *On the Nobility and Advantage of the Female Sex,* published by the St. Petersburg Academy of Sciences in 1784, has a political context. It can be seen not only as excessive in eulogizing women's bodily, spiritual, and moral dignities, but also as overstepping the limits, as sacrilegious. One should not miss, too, that glorifying women's various dignities happened against a background of the realities of "women's government,"[44] where the society knew about the sexual activities of the "mother of the Fatherland." Even so, the importance of judgments about women's dignity contains far-reaching, clear, and progressive opinions on women's role in all spheres of family and social life. Part of noble society, even in spite of Catherine II's adventures in her private life, positively welcomed her cultural, enlightened, and educational undertakings as Empress. The living conditions of the serfs in Russia allowed many privileges, even overstepping the limits of moral ideas about the permitted and the illicit in private life not only for landowners but also for their wives. And to them the private life of the "mother of the Fatherland" was the obvious case for imitation. So what is there to blame her and the treatise's author for? Perhaps only for his excessiveness. In this instance one can think that P. A. Alekseev, if he set as his objective to attract the attention of Catherine II to himself, achieved his goal. What can be more beautiful than to be punished by a woman for excessive love devoid of reasonable limits?

Who was the author of *On the Nobility and Advantage of the Female Sex*? It was not unusual for the eighteenth century that the translation of a work of art, drama, or science did not coincide with the original in everything. Today it could be called translation-alteration. Also, in St. Petersburg and Moscow, an expert sometimes translated not from the original language but from languages that served as mediators, usually from French. Besides this, translators very often put events and facts in the text to make it clearer to Russian audiences. In this new form, a work might become well known in Russia while it was forgotten in the land of the original author.

Iu. N. Solonin asks if Agrippa Nettesheim was really the author of the treatise. If he was, then how close to the original is the translation? Or was it a free exposition with some important generalizations and additions by the new author? Given the customs mentioned above, it is quite possible that the name of Agrippa Nettesheim, which was not even mentioned in the edition, like the name of the dead brother, sheltered the true author, the archpriest P. A. Alekseev. In any case confirmation of this requires additional research beyond the scope of this essay. What we want to emphasize is that the treatise appeared in Russian published by the St. Petersburg Academy of Sciences supervised by E. R. Dashkova, an active patroness of translation, and marked the growth of interest in the women's issue in Russian public society. And rumours about the displeasure of Catherine II and her retinue, though unconfirmed, could, in keeping with Russian tradition, only increase readers' interest in this work.

A supporter of the opinion that no one but P. A. Alekseev, archpriest of the Kremlin's Archangel Cathedral, was the author of the translation is A. N. Korsakov, a well-known researcher of Alekseev's personality and activities. Thanks to his efforts, letters and other documents from Alekseev's archive have been published. Korsakov explained Alekseev's publication of the treatise as the delusion of a simpleton: "In sending Dashkova such a work, which discusses the nobility and advantage of the female sex, he flattered the vanity of President of the Academy of Sciences and the Empress, and hoped, of course, that his translation would become known to Catherine II."[45] As for us, we do not think Alekseev was a simpleton. He acted as many others acted in his time who wanted to be noticed by the Empress, as others had acted in previous reigns. The archpriest might possibly know that Catherine II regarded with favor compliments addressed to her that extolled her as women's patroness.

Other examples confirm her satisfaction with being noted for her role as protectress of the female sex. We can refer to an ordinary example, an unimportant event, like the speech of a pupil of the Smolny Institute on the occasion of the ceremonial reception dedicated to the Empress's rewards in 1775. Naturally, to ensure that it included the words the Empress liked to hear, the speech was not prepared by the student herself. As was the custom, emphasis was put on the opening of Smolny Institute as the historical service of Catherine II, her special good deed for Russian women. The words of extravagant praise sounded solemn in the girl's mouth: "the great deed of the monarch, loving her people, who honored our sex, bereft of that useful knowledge which strengthens the mind and together with it reforms the heart, in the days of Your very glorious reign with those gifts."[46]

So should the highly experienced "simpleton," translator or author, archpriest of Archangel Cathedral, be, in his compliments, less extravagant than a girl from Smolny Institute? We are convinced that the point is not that contemporaries competed with each other with their brilliant and colourful panegyric toward the Empress. To praise a person sitting on the All-Russian throne in public like this was the norm, a norm that came into Catherine's century from previous reigns. Thus, we believe that P. A. Alekseev, supported by E. R. Dashkova, was responsible for the translation of Nettesheim's treatise and the radical opinions it contained.

## Notes

1. The author of this article has also researched the life and activity of Princess Dashkova, one of the closest associates of Catherine the Great. See G. A. Tishkin, "A Female Educationalist in the Age of the Enlightenment: Princess Dashkova and the University of St. Petersburg," in *History of Universities,* ed. Peter Denley (New York: Oxford University Press, 1994), 13:137–52; *Ekaterina Romanovna Dashkova: Issledovaniia i materialy* (St. Petersburg, 1996), T.8. Studiorum slavicorum monumenta: 80–94; Iu. D. Margolis and G. A. Tishkin, *Edinym vdokhnoveniem. Ocherkii istorii universitetskogo obrazovaniia v Peterburge v kontse XVIII-pervoi polovine XIX v.* (St. Petersburg: Izdatelstvo St.-Peterburgskogo universiteta, 2000), 227; G. A. Tishkin, ed., *Materialy po istorii Peterburgskogo universiteta. XVIII vek. Obzor dokumentov Sankt-Peterburgskogo filiala Arkhiva RAN* (St. Petersburg: Izdatelstvo St.-Peterburgskogo universiteta, 2000), 263.

2. G. R. Derzhavin, *Sochineniia* (St. Petersburg, 1866), iii. 621.

3. She had married Prince Mikhail Vorontsov, a colonel in the horseguards, at the age of fifteen.

4. The two academies filled separate functions. Members of the St. Petersburg Academy worked in the mathematical and natural sciences; members of the Russian Academy were primarily philologists.

5. V. V. Ogarkov, *E. R. Dashkova: ee zhizn' i obshchestvennaia deiatel'nost'* (St. Petersburg, 1893), 46–48.

6. *Modnoe ezhemesiachnoe izdanie, ili Biblioteka dlia damskogo tualeta,* 1779, pt. 2, 1117.

7. E.g.: Renal, Pokhvala Elize Draner, *Moskovskii zhurnal,* April 1792, pt. 6, 10–17.

8. N. M. Karamzin, "Portret milo zhenshchiny," *Vestnik Evropy,* January 1802, pt. 1, 55–59.

9. P. I. Makarov, "Nekotorye mysli izdatelia Merkuria," *Moskovskii Merkurii,* 1803, 1:4–18.

10. *Slovo o vospitanii, na vseradostnyi den' tezoimenitstva velikogo gosudaria, imperatora i samoderzhtsa vseia Rossii Pavla Pervogo, v torzhestvennom sobranii imp. Moskovskogo universiteta, iiunia 30 dnia 1798 goda, govorennoe nadvornym sovetnikom, professorom entsiklopedii i estestvennoi istorii, tsenzorom pechataemykh knig i glavnym smotritelem Blagorodnogo pri Universitete Pansiona, Antonom Prokopovichem-Antonskim* (Moscow: v Universitetskoi tipografii, 1798), 37.

11. S. E. Desnitskii, *Iuridicheskie rassuzhdeniia o nachale i proishozhdenii suprezhestva* . . . (Moscow, 1775), 27–29.

12. Iu. D. Margolis and G. A. Tishkin, *Otechestvu na pol'zu, a rossiianam vo slavu* (Leningrad, 1988), 168–69.

13. E. Shchepkina, *Iz istorii zhenskoi lichnosti v Rossii* (St. Petersburg, 1914), 146–97.

14. *Russkii Arkhiv.* 1871. Kn.9. P. 216–17. Pis'mo VI.

15. Dashkova, *Zapiski* (1987 ed.), 156. Obviously, numbers under Dashkova still remained very small. In the rest of Europe even the smallest university would expect to have one hundred students.

16. Lozinskaia, *Vo glave dvukh akademii,* 126.

17. Lozinskaia, *Vo glave dvukh akademii,* 126.

18. *Protokoly,* iii. 695, 720.

19. D. A. Tolstoy, "Akademicheskii universitet v 18 stoletii," *Zapiski Imperatorskoi Akademii Nauk* 51, suppl. 2 (St. Petersburg, 1885), 61.

20. Tolstoy, "Akademicheskii," 649. There is evidence that on 29 September 1783 a letter of thanks from Black was read (dated 4 September), but the full text of the original has not been preserved.

21. See *Sistematicheskii i alfavitnyi ukazatel' statei, pomeshchennykh v izdaniiakh i sbornikakh Imperatorskoi Akademii Nauk, a takzhe sochinenii, izdannykh Akademiei otdel'no, so vremeni ego osnovaniia 1872* g. (St. Petersburg, 1875), ii. 153, 351.

22. Ogarkov, *E. R. Dashkova,* 16.

23. M. I. Sukhomlinov, *Istoriia Rossiiskoi Akademii,* Vyp. 1 (St. Petersburg, 1874), 289–90.

24. See *O blagorodstve i preimushchestve zhenskogo pola. Iz istorii zhenskogo voprosa v Rossii,* ed. R. Sh. Ganelin (St. Petersburg, 1997), 18–30.

25. Iu. N. Solonin, *Golos o zhenskom dostoinstve iz XVIII veka: Iz istorii zhenskogo voprosa v Rossii,* ed. R. Sh. Ganelin (St. Petersburg, 1997).

26. V. V. Kolominov and M. Sh. Fainshtein, *Khram muz slovesnykh Iz istorii Rossiiskoi Akademii* (Leningrad, 1986), 143–45.

27. D. Sokolov, *Naznachenie zhenshchiny po ucheniiu slova bozhiia* (St. Petersburg, 1862), 8–9.

28. Genrikh Agrippa, *O blagorodstve i preimushchestve zhenskogo pola: Siia kniga perevedena v Moskve pod rukovodstvom Moskovskogo Arkhangel'skogo sobora protoireia Petra Alekseeva.* V St. Petersburg, izhdiveniem Imp.Ak.nauk 1784 goda, 2–3.

29. *Tserkovnyi vestnik* (1875), no. 2, 17.

30. Sv. Grigorii Nisskii, *Ob Ustroenii cheloveka,* ed. A. L. Verlinskii, trans. V. M. Lurje (St. Petersburg, 1995), 14–15; see also 7–8.

31. Solonin, *Golos o zhenskom dostoinstve iz XVIII veka,* 23.

32. Agrippa, 28.

33. Agrippa, 32–33.

34. Agrippa, 49.

35. M. M. Shcherbatov, "O povrezhdenii nravov v Rossii," in *O povrezhdenii nravov v Rossii kniazia Shcherbatova i puteshestvie A.Radishcheva* (Moscow, 1984).

36. Agrippa, 51.

37. Agrippa, 50.

38. Agrippa, 54

39. Agrippa, 62.

40. N. B., *Istoriia Rossiiskoi Akademii M. I. Sukhomlinova,* Vyp. 1 (St. Petersburg, 1874). Review in *Tserkovnyi vestnik* (1875), no. 2, 17.

41. Agrippa, 77, 78.

42. N. G. Chernyshevskii, *Polnoe sobranie sochinenii,* 11:273.

43. M. L. Mikhailov, *Vmeste: Delo* (1870), no. 1, 32.

44. See V. Mikhnevich, *Zhenskoe pravlenie i ego protivniki: Istoricheskii vestnik* (1882), bk. 7, nos. 2–3.

45. A. N. Korsakov, "P. Alekseev, protoirei Moskovskogo Arkhangel'skogo sobora," *Russkii Arkhiv* (1882), bk. 2, 175–76.

46. Cited from E. Likhacheva, *Materialy po istorii zhenskogo obrazovaniia v Rossii (1806–1856)* (St. Petersburg, 1899), 226.

# Suggested Readings

The authors of this collection hope that you will wish to explore the ideas and personalities presented in each of their case studies. They have written extensive notes to help you in your explorations. In addition, as editor, I would like to recommend some more general titles that can serve as your introduction to many of the analytical premises that underlie all of our studies: the questioning of the traditional linear narrative of the evolution of modern science; the omission in that story of women's contributions; the emergence of science as separate from philosophy in combination with the regendering of learning.

Scholars such as Alexander Koyré, a historian of science, and Ernst Cassirer, a philosopher, coined the terms "Scientific Revolution" and "Enlightenment" in the 1920s and 1930s and inaugurated what became the accepted narrative of these eras. Koryé gave his first lectures on the history of science in the 1930s. His approach became the basis of books like Herbert Butterfield's *The Origins of Modern Science, 1300–1800* (New York: Free Press, 1949) and A. Rupert Hall's *The Revolution in Science 1500–1750* (London: Longman, 1983; originally published in 1954 as *The Scientific Revolution, 1500–1800*), both widely reprinted in successive editions. See, for the traditional linear narrative presented historiographically, H. Floris Cohen, *The Scientific Revolution: A Historiographical Inquiry* (Chicago: University of Chicago Press, 1994). Ernst Cassirer's key text *The Philosophy of the Enlightenment,* published in 1932 and translated into English in 1951 (Princeton, NJ: Princeton University Press, 1951), became the classic philosophical analysis of eighteenth-century thought.

All of the essays in this collection challenge their interpretations. On the connotations of critiquing these older authorities for all historians of science and philosophy, on the need "to interrogate our own presuppositions . . . and how those presuppositions affect what we see in the past," see Margaret Osler, "The Canonical Imperative: Rethinking the Scientific Revolution" in *Rethinking the Scientific Revolution* (New York: Cambridge University Press, 2000).

A number of books give a picture of what women did in these centuries. Mary Ellen Waithe's *A History of Women Philosophers,* vol. 3, *Modern Women Philosophers, 1600–1900* (Boston: Kluwer, 1991) includes long sections on women like Conway and Du Châtelet. Margaret Atherton's *Women Philosophers of the Early Modern Period* (Indianapolis: Hackett, 1994) and Erica Harth's *Cartesian Women: Versions and Subversions of Rational Discourse in the Old Regime* (Ithaca: Cornell University Press, 1992) survey women's involvement in philosophical debates more specifically in the seventeenth and eighteenth centuries. Londa Schiebinger's *The Mind Has No Sex? Women in the Origins of Modern Science* (Cambridge, MA: Harvard University Press, 1989) tells of women engaged in the scientific debates and activities of their day. For a present-oriented summary of this phenomenon, see also Schiebinger's *Has Feminism Changed Science?* (Cambridge, MA: Harvard University Press, 1999), in which she comments on the fact that "the exclusion of women was not a foregone conclusion" but, instead, was coincident with the "stringent formalization of science in the nineteenth century."

Excellent collections exist that describe various aspects of the gradual separation of "science" from "philosophy." Margaret J. Osler's *Rethinking the Scientific Revolution* (New York: Cambridge University Press, 2000) presents key essays on the historiography of this evolution. Roy Porter and Mikulás Teich's collections *The Scientific Revolution in National Context* (New York: Cambridge University Press, 1992) and *The Enlightenment in National Context* (New York: Cambridge University Press, 1981) present monographic studies in a series of national settings. The essays in William Clark, Jan Golinski, and Simon Schaffer's *The Sciences in Enlightened Europe* (Chicago: University of Chicago Press, 1999) do the same for the theory and practice of "science" across Europe during the era of the Enlightenment. On the ways this separation evolved in the works of specific individuals, see, for example, Wolfgang Lefèvre, ed., *Between Leibniz, Newton, and Kant: Philosophy and Science in the Eighteenth Century* (Boston: Kluwer, 2001).

Our modern concept of science arose out of discussions of subjectivity, certainty, rationality, and probability in early modern natural philosophy. To understand the complexity of these issues, see, for example, Susan James, *Passion and Action: The Emotions in Seventeenth-Century Philosophy* (Oxford: Clarendon Press, 1997); Barbara J. Shapiro, *Probability and Certainty in Seventeenth-Century England: A Study of the Relationships between Natural Science, Religion, History, Law, and Literature* (Princeton, NJ: Princeton University Press, 1983); and Lorraine Daston, *Classical Probability in the Enlightenment* (Princeton, NJ: Princeton University Press, 1988). Experiments seemed the way to create "facts," but even this practice had no fixed meaning. Two essays in Peter Dear's collection *The Scientific Enterprise in Early Modern Europe: Readings from Isis* (Chicago: University of Chicago Press, 1997) offer particularly good explanations of the problems. Keith Hutchinson describes the turn away from the metaphysical in "What Happened to the Occult in the Scientific Revolution?" and Steven Shapin explores the household practice of science in "The House of Experiment in Seventeenth-Century England." For more recent questioning of the nature of "experiment," see *The Uses of Experiment: Studies in the Natural Sciences,* ed. David Gooding, Trevor Pinch, and Simon Schaffer (New York: Cambridge University Press, 1993).

Philosophy as it emerged in this transition era from the seventeenth century to the mid-nineteenth century also shifted goals: from the study of "truth claims" to the study of "process" and from "linguistic, historical, political and psychological inquiries into forms of knowledge construction and conflicts within particular discursive formations." Thus the shifts are described by Jane Flax in her essay "The End of Innocence," in *Feminists Theorize the Political,* ed. Judith Butler and Joan W. Scott (New York: Routledge, 1992).

The term "scientist" was coined at the 1833 meeting of the British Association for the Advancement of Science to suggest professionalism, but this shaping of new terms had its gendered consequences. In that context the word was assumed to designate a man. Erica Harth *(Cartesian Women)* describes the traditional "uneasiness of the relationship between educated women and rational discourse." The ancient images of women as incapable of reason had never really been superceded. On the connections between language and the actual configuration of science as a male domain, see Ludmilla J. Jordanova's introduction to *Languages of Nature: Critical Essays on Science and Literature* (London: Free Association Books, 1986) and her *Sexual Visions: Images of Gender in Science and Medicine between the Eighteenth and Twentieth Centuries* (Madison: University of Wisconsin Press, 1989). For discussion of these subtleties in a more modern context, see Donna J. Haraway, "Modest Witness: Feminist Diffractions in Science Studies," in *The Disunity of Science: Boundaries, Contexts and Power,* ed. Peter Galison and David J. Stump (Stanford, CA: Stanford University Press, 1996).

As you can see, this new approach to the histories of science and philosophy offers exciting challenges and the potential for new discoveries. It documents the contributions of women. It makes evident the intellectual and social consequences of this key reshaping of knowledge and the methods for its validation. Moreover, it demonstrates the significance of gender, however unintentional, in the creation of our modern-day institutional authorities.

# Index

# List of Contributors

**Susanna Åkerman** has been a research fellow in the Department of the History of Ideas, Uppsala University, and in the Department of the History of Ideas at Stockholm University. She is currently at the Swedenborg Library in Stockholm. Published in both Swedish and English, her scholarship has highlighted the eclectic characteristics of "natural philosophy" in the seventeenth century and Queen Christina of Sweden's active engagement in what today would be characterized as both scientific and metaphysical speculation. As the author of *Rose Cross over the Baltic: The Spread of Rosicruciansism in Northern Europe* (1998), Åkerman is a leading expert on this period of northern European intellectual history. She wrote the biographical study *Queen Christina of Sweden and Her Circle: The Transformation of a Seventeenth-Century Philosophical Libertine* (1991) and has spoken widely on her research in the United States and Europe.

**Franco Arato** is known in Italy for his work on the Venetian Francesco Algarotti's association with Voltaire and his popularization of Newtonian experiments and theories about optics and universal attraction, *Il Newtonianismo per le dame (Newton's Philosophy for the Ladies)*. He is the author of *Il secolo delle cose: Scienza e storia in Francesco Algarotti* (1991); *Letterati e eruditi tra Sei e Setecento* (1996); and *La storiografia letteraria nel Settecento italiano* (2002). He is a regular contributor to *Belfagor* and *Giornale storico della letteratura italiana.*

**Stephen Clucas**, Senior Lecturer in English and Humanities at Birkbeck, University of London, has written widely on the history of sixteenth- and seventeenth-century "science" and "philosophy." His publications include a translation of Paolo Rossi's historical study *Logic and the Art of Memory: The Quest for a Universal Language* (2000) and most recently an edited collection of essays on the seventeenth-century poet, playwright, and natural philosopher Margaret Cavendish, *A Princely Brave Woman: Essays on Margaret Cavendish, Duchess of Newcastle* (2002).

**Lynette Hunter** is Professor of Rhetoric at the University of Leeds and Professor of the History of Rhetoric and Performance at the University of California, Davis. She is the author of several monographs on rhetoric in science, literature, and computing. In her book *Critiques of Knowing: Situated Textualities in Science, Computing, and the Arts* (1999), for example, she analyzes the rhetoric of modern science and shows its relationship to the rhetoric of liberal politics. In her work, Hunter focuses on the place of women as writers, readers, and performers. Her essay in this collection arises from her interest in women's household science. She is currently engaged in a study of rhetorics used by communities new to democracy around the world.

**Monika Mommertz**, Professor of Early Modern History at the Humboldt University of Berlin, in explaining "household production" of astronomical and astrological information by two generations of women of the Winkelmann-Kirch family, continues to focus her scholarship on what she describes as the "shadow economy" of scientific work. Mommertz is known in Europe for her original insights on the role of the household and the institutional strategies that rendered women and their scientific contributions "invisible."

**Margaret J. Osler** is a Professor in the Department of History at the University of Calgary. Her essays have appeared in leading North American and British historical journals. In a number of her publications, most notably *Divine Will and the Mechanical Philosophy: Gassendi and Descartes on Contingency and Necessity in the Created World* (1994), Osler explores the relationship between theology and natural philosophy in the early modern period and challenges the progressive view of the history of science.

**J. B. Shank**, Assistant Professor of History at the University of Minnesota, has been named a McKnight Land-Grant Professor for 2004–2006, a special award honoring the most promising junior faculty at the university. His book *Before Voltaire: Newton, "Newtonianism," and the Origins of the Enlightenment in France,* which examines the cultural history of Newton's reception in France between 1687 and 1750, is forthcoming from the University of Chicago Press. His articles on the history of early modern French science have appeared in scholarly journals, and he has spoken at numerous conferences on European cultural and intellectual history and on the history of science.

**Hilda L. Smith**, in her most recent book, *All Men and Both Sexes: Gender, Politics, and the False Universal in England, 1640-1832* (2002), continues her exploration of men's and women's participation in the key intellectual and political developments of England. She is author of *Reason's Disciples: Seventeenth-Century English Feminists* (1982) and the co-compiler of the award-winning annotated bibliography *Women and the Literature of the Seventeenth Century.* Her essay in this volume forms part of a larger exploration of the works of the English prodigy Margaret Cavendish, Duchess of Newcastle. Smith is Professor of History at the University of Cincinnati.

**Grigory A. Tishkin** is best known in Russia for his scholarship on the University of St. Petersburg, for which he wrote the official history. He has published widely in Russia on the nineteenth-century Russian women's movement and on the history of university education. His article "A Female Educationalist in the Age of the Enlightenment: Princess Dashkova and the University of St. Petersburg" appeared in an English collection on the history of universities for Oxford University Press in 1994.

**Judith P. Zinsser's** essay on the marquise Du Châtelet is part of a larger project, a biography of this multitalented participant in the first decades of the French Enlightenment, forthcoming from Viking/Penguin (2006). She is also collaborating on a translation of Du Châtelet's selected works for the University of Chicago series The Other Voice in Early Modern Europe. Zinsser's essays on the marquise and her writings have appeared in *French Historical Studies, SVEC, Notes and Records of the Royal Society,* and in the edited collection *Women Writers and the Early Modern British Political Tradition* (1998). She is coauthor of *A History of Their Own: Women in Europe from Prehistory to the Present* (2000). Zinsser is Professor of History and affiliate in Women's Studies at Miami University (Ohio).